SUSTAINABILITY EDUCATION FOR CHILDREN AND YOUNG PEOPLE

This seminal volume responds to the pressing need to prepare all children and young people for a sustainable future in light of the climate crisis, providing clear and accessible information and strategies on how to fully embed sustainability into pedagogy and supporting current and future educators.

This book explores key environmental issues before exploring a range of pedagogies, each grounded in hope and exemplifying the principles of inclusion, empowerment, agency and social justice. Chapters explore major themes such as climate migration and displacement, climate activism, nature connectedness, sustainable leadership and eco-anxiety to offer practical, scalable pedagogies that connect with the natural world. Designed to channel climate anxieties into energy for a reinvigorated, sustainable future of education, this book encourages an accessible and considered approach to classroom practice. Featuring contributions from wealth of international experts in the field, the volume ultimately provides a valuable resource for educators and students of education in all phases to develop the confidence to take an active lead in shaping education to help ensure a sustainable future.

This book will appeal to educators, policymakers and researchers in environmental and sustainability education, teacher education, social justice education and educational leadership. Curriculum designers as well as those working across primary, secondary and higher education will also find the volume of interest.

Nicola Watson has taught in the early years, primary and higher education sectors for over twenty years, and most recently at the University of Worcester.

Elena Lengthorn is a Senior Lecturer in Teacher Education, University of Worcester, UK, where she leads the PGCE Geography programme and Education for Sustainable Futures Special Interest Group and Chairs the Sustainability Forum.

Carla Solvason is a Senior Lecturer in the Department for Children and Families, University of Worcester, UK, where she leads the Social Pedagogy Research Group.

Geoffrey Elliott is Professor of Post-Compulsory Education at the University of Worcester, UK.

SUSTAINABILITY EDUCATION FOR CHILDREN AND YOUNG PEOPLE

Educating for the Environmental Crisis

Edited by
Nicola Watson, Elena Lengthorn,
Carla Solvason and Geoffrey Elliott

Routledge
Taylor & Francis Group
LONDON AND NEW YORK

Designed cover image: © Getty Images

First published 2025
by Routledge
4 Park Square, Milton Park, Abingdon, Oxon OX14 4RN

and by Routledge
605 Third Avenue, New York, NY 10158

Routledge is an imprint of the Taylor & Francis Group, an informa business

British Library Cataloguing-in-Publication Data
A catalogue record for this book is available from the British Library

ISBN: 978-1-032-82894-7 (hbk)
ISBN: 978-1-032-79635-2 (pbk)
ISBN: 978-1-003-50685-0 (ebk)

DOI: 10.4324/9781003506850

Typeset in Sabon
by codeMantra

This book is dedicated to August Jean Knight.
May you and your generation share in the wonder of this world

CONTENTS

EDITORS AND CONTRIBUTORS

Kathleen Bailey is a Doctoral Researcher and Tutor at Sussex University. Working with children and young people, she advocates for a posthumanist approach towards outdoor pedagogy that holds equity and sustainability at its heart.

Gareth Dart is Course Leader for the MA SEN/ALN at the University of South Wales. Having previously worked in Botswana, he maintains an interest and involvement in inclusive education in the context of resource-poor countries.

Geoffrey Elliott is Professor of Post-Compulsory Education at the University of Worcester, having taught in comprehensive schools, further, adult and higher education. He edits the international peer-reviewed journal *Research in Post-Compulsory Education*.

Antoinette Frearson is a Doctoral Researcher at the University of Worcester. As well as sustainability and climate change education, Antoinette's research interests also include childhood adversity, trauma-informed practice and SEND.

Janet Harvell is a Senior Lecturer at the University of Worcester. She has worked in education for 35 years. Her interests focus on international approaches, and this has included research in China and with refugee families in Calais. She also works with charities promoting the adoption of eco-friendly practices.

Rosemarie Hill is a Lecturer in Primary Education at the University of Worcester. She specialises in primary history and English. Previously, she has taught in primary schools for 15 years.

Dan Hughes is a Principal Lecturer in Primary Education at the University of Worcester and is the PGCE Primary Course Leader. He specialises in PSHE and R(S)HE, English and PE. He has previously taught for 14 years in primary school.

Sam Jackson is Director of Climate Science & Impact at Ecologi, a social enterprise that helps individuals, families and businesses reduce their carbon footprint and take climate action.

Perpetua Kirby is a Lecturer in Childhood & Youth Studies at the University of Sussex. She convenes the course and teaches a module creating a Forest Food Garden on the university campus. She co-leads a research programme on sustainability education.

Emma Laurence is a Lecturer in the Department for Children and Families at the University of Worcester and Course Leader of the BA (Hons) Early Childhood in Society.

Elena Lengthorn is a Senior Lecturer at the University of Worcester where she leads on PGCE Geography and Education for Sustainable Futures Special Interest Group. She was awarded 'Global Educator of the Year' in 2016. She is the Co-Chair of the Our Shared World network and an advisor on the OCR Natural History Strategic Advisory Board.

Shaun McInerney taught and led schools in the UK and India for 25 years. Working at the University of Worcester, he uses coaching and leadership development to support education leaders unlock potential and respond to the needs of young people in a rapidly changing world.

Ella Partington graduated with a first-class honours degree in Education Studies from the University of Worcester in 2024. She became involved in writing this chapter by choosing to be a co-author as an assignment for her 'Radical Education' module.

Morgan Phillips is the Director of Education at the environmental charity Global Action Plan. He has previously held roles as Head of Eco-Schools England and Co-Director of climate change adaptation charity The Glacier Trust.

Alison Prowle is a Senior Lecturer at the University of Worcester, where she specialises in teaching and research connected to trauma-aware practice and children's adverse life experiences. She has a particular interest in climate migration.

Lowri Prowle is a Psychological Wellbeing Practitioner, who works for an NHS Talking Therapies Service. Lowri is passionate about supporting children and young people's mental health and volunteers as a young people's advocate and mentor.

Tamzin Ractliffe is a Social Entrepreneur and the Founding Director of Impact Trust. She led the creation of the accredited Routes2Resilience programme to provide transformative experiential learning spaces for new ways of thinking and doing education in sustainability and resilience leadership.

Carla Solvason was a Primary School Teacher before moving into Higher Education, where her chief responsibility is research. She leads the Social Pedagogy Research Group at the University of Worcester and is passionate about research that has a positive social impact.

Samantha Sutton-Tsang is a Senior Lecturer and Course Leader for the FdA Early Years at the University of Worcester. With extensive experience in early childhood education, Samantha is dedicated to fostering innovative teaching practices and supporting professional development.

Nicola Watson began her career as a Family Lawyer and Mediator. She then taught in the early years, primary and higher education sectors for over 20 years, and most recently as a Senior Lecturer at the University of Worcester. She is committed to promoting compassion and care for the human and the more-than-human world.

Rebecca Webb is a Senior Lecturer in Education at the University of Sussex. She has convened the MA in Early Years Education and the MA in Education. She has worked as a teacher and local education authority advisor for Equality, Diversity, Achievement and Inclusion.

Vicky Whitworth is a Co-Founder of Chardstock Eco Group. Current projects include working with the local primary school on a range of projects. The group was a finalist in the BBC Radio Devon Make a Different Award 2024.

FOREWORD

With good reason, climate change is often described as the greatest challenge to have faced humanity since the dawn of time. The impact of the Industrial Revolution in the mid-19th century led to humans releasing enormous quantities of greenhouse gases (GHG) into the Earth's atmosphere. This was through the burning of fossil fuels for the production of energy, for mass transportation of goods and people and for the over-consumption of the Earth's scarce resources, particularly by economically developed countries. These greenhouse gases prevent heat energy from leaving the Earth with the result that the Earth's temperature is rising on land, in the oceans and in the atmosphere. This is the driver of climate change.

What is clear is that the battle to stop climate change and protect the sustainability of the Earth is not an objective that can easily be achieved. The UN Framework Convention on Climate Change (UNFCCC) has set climate change objectives for the world, one of which involves constraining global warming to 1.5°C. Climate scientists strongly suggest that even an increase of 1.5° will have calamitous impacts with larger temperature rises having much greater consequences. Tackling this has to involve the many and not the few. It requires contributions from all countries, from all groups and from all individuals on the planet – everyone has a part to play. Unfortunately, at the time of writing, the situation doesn't look good, and the 1.5° temperature rise looks set to take place in the next few years, with potentially higher temperatures to follow. The consequences are likely to be catastrophic, particularly in developing countries.

A key focus in stopping climate change and maintaining sustainability must be to influence the views and actions of younger generations across the world. While older generations have a part to play, and also need to take similar actions, the reality is that it is younger generations who will face the realities and difficulties of ongoing climate change that will take place decades into the future. Hence, they have a huge personal stake in maintaining a sustainable planet. Also, it might be argued that younger generations are more open to change than older generations. Hence, it is vital to educate them to understand: the dangers of ongoing climate change, its implications and the actions that need to be taken to mitigate these problems and avoid disaster.

In the light of this extremely worrying situation and the need to focus on younger generations, it is wonderful to see the publication of this book *Sustainability Education for Children and Young People: Educating for the Environmental Crisis*. This book is comprehensive in nature and has three important features.

First, it covers all parts of the education system from early years through to university level. Second, it covers a broad range of educational topics including children's rights, curriculum design, education leadership and holistic education. Finally, and perhaps most importantly, it focuses on a number of specific topics that are vitally important in sustainability education such as recognising children's rights, developing curiosity about sustainability, promoting social justice and dealing with eco-anxiety.

This is a book that must be on the bookshelf of every education practitioner who is interested in sustainability but also those practitioners who, currently, are not focused on sustainability but need to be.

I can thoroughly recommend this book.

Professor Malcolm J. Prowle
Author of *Global Climate Change: A realistic view of the future* and *Global Climate Change: A guide for future action*
November 2024

INTRODUCTION

Nicola Watson and Elena Lengthorn

Sustainability: the critical role of educators

As these words are being written, it's a dull day in England and through the window, the leaves can be seen, changing from shades of green to red, orange, brown and gold. The sky is grey and the air misty. It's mild but cold enough to warrant an extra layer, a reminder that summer has transitioned to autumn, as it has for millennia. It is reassuring for those of us who live here and now to experience the continuing cycle of the environment upon which all life depends. At this time, world leaders are meeting for the COP16 UN biodiversity conference in Columbia. The conference is billed as a united attempt to halt the destruction of the planet. They are discussing plans which include protecting 30% of the planet for nature by 2030. This was one of a range of targets agreed at COP15 in 2022 by almost every country in the world. This suggests that although climate change and biodiversity loss are major concerns, we can be assured that our political leaders are attending to what needs to be done on our behalf. However, more than 80% of countries failed to submit plans to meet the agreement reached at COP 15 (IPBES, 2024). This is no time for complacency. Shue (2022, p. 5) states that:

> Failure to act urgently now is a dire threat to future generations. If delay now allows the passing of physical or social tipping points…we miss our unparalleled but fleeting opportunity to make life better for our grandchildren and instead inflict a legacy of danger.

Educators play a crucial role in responding to the environmental crisis, a conclusion that has been recognised at the highest levels of UK Government

DOI: 10.4324/9781003506850-1

with the former Secretary of State for Education, the Rt Hon Nadim Zahawi, introducing their Sustainability and Climate Change Strategy with an acknowledgement of the formidable challenge of climate change and the need for learners to "know the truth about climate change" and to be given "the hope that they can be agents of change through hands on activity..." (DfE, 2023). Those who work with children and young people (CaYP) are charged with caring for, developing an equipping them for the world in which they live and which future generations will inherit. This applies not only to teachers, but to all who work in the diverse contexts of education: from nursery nurses, to youth workers, to university professors. We use the abbreviation 'CaYP' throughout this book to reflect the wide range of those connected with education in all its forms. As Armon et al. (2020, p. 15) point out, "One of the features of sustainability education is that while it does not command a discrete body of knowledge, it actually touches on just about everything."

Sustainable development is a contested term. We do not intend to engage in philosophical debate here but instead have adopted the definition agreed in the 1980s, by the World Commission on Environment and Development (WCED) as "development that meets the needs of the present without compromising the ability of future generations to meet their own needs" (Brundtland Report, WCED, 1987), with the sustainable development of CaYP being actively delivered by their families and educators.

The aims of this book

Warnings of an uncertain future may make us respond with feelings of fear, anxiety and powerlessness. As human beings it's a natural response to defend ourselves against perceived threats. However, ignorance or denial will not serve us or the CaYP with whom we work. Instead, we need to acknowledge and reflect on our feelings, recognise our own defences and work with them constructively. We can connect with others to share values and goals to bring about individual and social change (Adams, 2018). This constructive approach supports our own mental health and helps us support others appropriately.

The central aim of this book is to offer hope to those who work with and educate CaYP and to support the development of eco-centric orientation and practice. Ypi (2023) makes a powerful statement when she asserts that "Being hopeful is not about guaranteeing the right outcome but preserving the right principle: the principle based on which a moral world makes sense." Bourn (2021) suggests that many young people are feeling ignored, overlooked, powerless and anxious, with young people asking whether or not they should bring children into the world. Our book offers ways in which we might help them move from feeling helpless into individual and collaborative action.

We all need to harness knowledge, courage and resolve to develop as agents of positive change. Most educators recognise the need to shift the focus in education towards understanding the environmental crisis. A 2019 poll, undertaken by Oxfam on the climate crisis, reported that 69% of UK teachers wished to see more teaching in schools about climate change, and 70% supported radical change in UK legislation to make the education system fit for the times in which we live (Howard-Jones et al., 2021). That is not to suggest that the way ahead is simple. Philosopher of education, Zachary Stein (2019, pp. 8–9) suggests that the environmental crisis creates a meta-crisis in education. He identifies four elements of the meta-crises of sense-making, capability, legitimacy and meaning. These elements can result in:

- confusion and difficulty in comprehending what is true in an increasingly complex world (sense-making)
- an inability to solve problems at the level needed (capability)
- questioning where responsibilities lie, professionally, culturally and politically (legitimacy)
- asking "What am I doing and why? Does my professional role align with my own experience and values and what I believe to be true?" (meaning).

The collective responses to the 'meta-crisis' explored in this book aim to support educators to make sense of their role within their own unique contexts, develop capability, feelings of legitimacy and enhance meaning and purpose to their roles.

In his article on the tactics of delay as a form of climate change denial, Shue (2022) concludes that it would be a mistake to assume that all measures taken must be extreme. Indeed, he suggests that what's needed from our political leaders is patient coalition building, imagination and a willingness to be unconventional to get results. As educators, we want to engender understanding, moral consistency, imagination and curiosity in the CaYP with whom we work, developing their sense of being valuable and the belief that their individual actions, matter. We need to begin by developing these dispositions ourselves. It is crucial that we work together, rather than in isolation. While we all have choices and agency to some degree; our contexts determine our opportunities for change making. That is why curiosity and imagination are powerful drivers of positive change. Forming connections and networks of mutual support with our students, colleagues, friends and communities wherever opportunities arise will enable knowledge creation and relieve the unhelpful pressure of individual responsibility and perhaps burnout. Eco-anxiety affects educators as well as CaYP and taking positive action, particularly with others can help provide hope and meaning in challenging times.

The themes of this book

The foundation of hope upon which this book sits is underpinned by four themes which are woven throughout the chapters. They comprise inclusion, empowerment, agency and social justice. As ethical principles they help to identify and act upon opportunities to make a positive contribution through our work:

Inclusion

Inclusion demands that we look for, listen to and respect the voices of those who are marginalised and vulnerable to inequalities of wealth and power, including future generations who as yet, have no voice. They will be our future CaYP and will inherit the legacy which current generations leave. It is illuminating to engage in a thought experiment; asking ourselves what we would say to them and how we would explain our society's part in the legacy we leave.

Social justice

Social justice necessitates the recognition that for many CaYP around the world, the climate crisis is here and now and this is reflected throughout this book which includes analysis and case studies from around the world. As educators we will encounter students who have suffered directly from the effects of the environmental crisis. In wealthy countries where resources are spread unequally, many CaYP are deprived of access to clean air, nature-rich outdoor spaces and vulnerability to flooding. In poorer countries poverty, war and lack of access to life-sustaining resources are all exacerbated by climate change. We contend that a holistic view of the world is fundamental to addressing the problems we face. As a species, our interests are inextricably bound with one another and with all other species which together comprise Earth's ecosystem. Fostering compassion and cooperation helps us connect with others, human and more than human and thus leads the way to repair and healing of our damaged world.

Empowerment

Empowerment requires a recognition that CaYP have a right to be informed about their world and their futures. We have a social and moral responsibility to ensure that they are prepared to speak out and make decisions about the world that they wish to inhabit. The issue of eco-anxiety as a potentially disempowering phenomenon is featured as a separate chapter, but this book as a whole is intended to help educators create the conditions to harness and amplify the creativity and energy of CaYP.

Agency

Agency can be defined as the ability and opportunity to carry out intentions. Educators who optimise their students' agency help support their desire to generate and to create, as the human mind has evolved to do (Bandura, 2001, p. 4).

Active participation enhances wellbeing and enables CaYP to locate their own, unique perspectives and find purpose and meaning (Watson, 2024).

This book captures the voices of experienced educators from early years to higher education, indigenous voices, academics, teachers, researchers and students. They work in a wide variety of contexts and throughout, ideas, suggestions and case studies are designed to stimulate educators' own ideas about opportunities in their own unique contexts. Our sincere hope is that in sharing our personal and professional experiences, grounded in a commitment to help build a better future, we can support educators in exploring ways of knowing, questioning and reimagining education. There is no more important a role than caring for CaYP and helping them develop the knowledge, skills and confidence to live well and sustainably so that future generations might do the same.

References

Adams, M. (2018) *Climate change, sustainability & defence mechanisms*. Available at: https://www.researchgate.net/publication/322701007_Climate_change_sustainability_defence_mechanisms.

Armon, J., Scoffham, S. and Armon, C. (2020) *Prioritizing sustainability education: A comprehensive approach*. London: Routledge, Taylor & Francis Group.

Bandura, A. (2001) 'Social cognitive theory: An agentic perspective', *Annual Review of Psychology*, 52(1), pp. 1–26. Available at: https://doi.org/10.1146/annurev.psych.52.1.1.

Bourn, D. (2021) 'Pedagogy of hope: Global learning and the future of education', *International Journal of Development Education and Global Learning*, 13(2), pp. 65–78.

Brundtland, G.H. (1987) *Our common future: Report of the world commission on environment and development*. Geneva, UN-Dokument A/42/427.

Department for Education (DfE, 2023) *Sustainability and climate change: A strategy for the education & children's services systems*. DFE Available at: https://www.gov.uk/government/publications/sustainability-and-climate-change-strategy (Accessed 20 November 2024).

Howard-Jones, P., Sands, D., Dillon, J., and Fenton-Jones, F. (2021) 'The views of teachers in England on an action-oriented climate change curriculum', *Environmental Education Research*, 27(11), pp. 1660–1680. https://doi.org/10.1080/13504622.2021.1937576.

Intergovernmental Science-Policy Platform on Biodiversity and Ecosystem Services (IPBES) (2024) *Global assessment report on biodiversity and ecosystem services*. Available at: https://www.ipbes.net/global-assessment (Accessed 26 November 2024).

Shue, H. (2022) 'Unseen urgency: Delay as the new denial', *WIREs Climate Change* [Preprint]. Available at: https://doi.org/10.1002/wcc.809 (Accessed 11 November 2024).

Stein, Z. (2019) *Education is the Metacrisis: Why it's time to see planetary crises as a species-wide learning opportunity*. London: Perspectiva. Available at: https://systems-souls-society.com/wp-content/uploads/2022/01/Education-is-the-metacrisis.pdf.

Watson, N. (2024) 'The value and benefits of creativity', in Taylor, S. *Creativity in the early years: Engaging children aged 0-5*. London: Sage, pp. 21–32.
Ypi, L. (2023) The day my students stopped me in my tracks – and changed how I thought about hope. *The Guardian*. Available at: https://www.theguardian.com/commentisfree/2023/dec/21/students-hope-change-enlightenment-world-worse?fbclid=IwAR1P8GFGBgTl_B0s9KY-nS6KmUt9Zds7eln2IHWcDEnI1CyYVdsi5- (Accessed 20 October 2024).

PART 1

The environmental crisis – why it matters to everyone

Nicola Watson

Humanity's predicament is complex and it is unrealistic to expect all professionals who work with children and young people to attain a specialist knowledge of all of the factors which drive the environmental crisis. However, it is imperative that all young people are given the opportunity to develop environmental literacy. Accordingly, the main aim of Part 1 is to provide an outline of the impact of industrialised societies on Earth's ecology, focusing on specific strands of concern. The section comprises a single chapter so as to reflect the nature of the environmental crisis; that each and every aspect is interconnected.

The starting point for embedding sustainability into our work must be to define our terms of reference. What exactly do we mean when we speak of the 'environmental crisis'? Although there are multiple perspectives on this (as reflected in the following sections), we begin by exploring and acknowledging the evidence. Human activity has impacts that are not always apparent, and sometimes, short-term economic interests prevail over long-term environmental security. False claims of environmental credentials might be made (termed 'greenwashing') or inconvenient truths might be buried, such as by the giants of the fossil fuel and animal agriculture industries.

Although denial of climate change and biodiversity loss is ubiquitous and is fuelled by misunderstanding, fear and confusion, the scientific community is almost unanimous that the climate changing is a direct consequence of human activity. We know that this is accelerating biodiversity loss which is in steep decline and that air, water and land are being increasingly adulterated by pollution. Human beings are animals, though this fact is often unacknowledged at an individual and cultural level, and like all life on Earth, we are dependent upon the elements to sustain us. Without them we cannot live.

DOI: 10.4324/9781003506850-2

Like all living creatures on Earth, we are remarkable. However, among the seemingly unique traits of humans we have the capacity to use complex and abstract language, to develop a moral sense and to be aware of our mortality. This means that we have the capacity to understand the consequences of our actions, a most powerful tool in combating the greatest challenge of our age, the environmental crisis.

1

WHAT IS THE ENVIRONMENTAL CRISIS AND WHY DOES IT MATTER?

Antoinette Frearson and Janet Harvell

Introduction: what is climate change?

The term 'climate' refers to regional atmospheric weather patterns such as temperature, wind, sun and rain over a particular area; typically averaged over 30 years. Although we can still experience general extremes in weather, such as a very wet spring, a particularly cold winter or a long summer drought, this does not necessarily mean the climate is changing. However, if ongoing measurements and analyses of these show that the conditions are prevailing across decades, this alerts us to potential climate changes occurring (British Geological Survey, 2023).

Climate change is not new. Fossils and rocks indicate that significant climate fluctuations have occurred several times in the Earth's past. For example, Earth has undergone four major ice-ages in the last 500,000 years (National Geographic, 2022). Similarly, fossilised plants from the Cretaceous period (144–165 million years ago), that grow in much warmer climates, have been found in Greenland demonstrating that the Earth was much warmer then, compared to present day (National Geographic, 2022). This illustrates how climate changes, up to now, have been natural rather than as a result of anthropogenic (human) activity, sometimes due to major volcanic eruptions, meteorites or variations in the solar activity. However, current changes cannot be explained by natural causes. After decades of careful measurements and analyses of longer-term changes in weather patterns and temperatures, scientists have identified that Earth is heating again, but this time, due to human rather than natural activity (BGS, 2023). Since the dawn of the Industrial Revolution, in the 1800s, the burning of fossil fuels such as coal, oil and

DOI: 10.4324/9781003506850-3

gas, as well as intensified agricultural methods, have been major drivers of greenhouse gas emissions and therefore, recent climate change (UN, 2023).

What is biodiversity and why is it important?

The term 'biodiversity' refers to all living things on Earth; from bacteria, flora and fauna, everything that lives underground, in the sea and air, as well as humans. Biodiversity is short for 'biological diversity' and is the name we give to the variety and volume of all life on Earth, how we all interact, support, and enable each other and the physical world. All living things have their own habitats and populations or ecosystems. Examples of these include our rivers (freshwater), seas and oceans (saltwater), deserts, forests, tundra, icecaps and urban areas. Although humans are often assumed to be independent of and in command of all life on Earth, we are indivisibly entwined in the natural world and rely on Earth's biodiversity to survive. It provides clean air to breathe, food to eat and fresh water to drink. In order to thrive, human beings and much else besides, need a natural world that is complex, resilient and full of variety. The benefits of this are known as 'ecosystem services' or 'natural capital,' all of which depend on a healthy environment (NatureScot, 2023). Examples of these ecosystem services include coastlands, which protect us from storm surges and changing sea levels, wetlands that keep pollution levels low, forests that break the flow of water and help prevent floods as well as human-made green infrastructures which have been recognised in helping reduce our anxieties and promote mental wellbeing (Natural England, 2023). Habitats may cope with small changes which have minimal effect, but bigger changes threaten the delicate balance of diversity (National History Museum (NHM), 2023).

Threatened species and mass extinctions

A loss of biodiversity accelerates extinction rates and in Earth's long history a number of mass extinctions have taken place, with millions of species dying out over time. In fact, 99% of all organisms that have ever lived are now extinct, so our planetary biodiversity has changed many times. There have been five mass extinctions in the past, one of these being 66 million years ago when dinosaurs became extinct. Some experts believe that we are in the sixth mass extinction, the Holocene Extinction (United Nations Environment Program, 2019), driven primarily by a growing human population and our unsustainable use of land, energy and water. It has been suggested that the species richness in more than half of our land ecosystems is now dangerously low and is compromising biodiversity's contribution to ecosystem services (NHM, 2023). Although this has happened naturally in the past, it is argued that humans are now causing more and more species to face an unpredictable future through deforestation for logging, farming and mining (NHM, 2022). There are two further crucial human activities that are driving this depletion of species: agricultural practices and pollution.

Farming

A recent report by the United Nations (UN, 2023) states that around one million species are at risk of extinction, with agriculture being stated as the main driver. This is reinforced by the UK State of Nature report (National Biodiversity Network, 2019) which observed that in the last 45 years, agriculture has been responsible for a 41% decrease in UK wildlife, and a 54% reduction in the Farmland Bird Indicator (a structural and sustainable development indicator by the European Union).

Despite total utilised agricultural areas in the UK decreasing by 2.3% over the last 20 years, agriculture still makes up 70% of the UK's total land-use (Department for Environment, Food and Rural Affairs (DEFRA), 2023), which makes farming one of the main drivers behind biodiversity loss and habitat destruction (from converting non-farming land into land used for agricultural purposes). This is overwhelmingly due to animal agriculture as most crops are grown to be fed to farmed animals. Furthermore, the intensification of farming, because of the increasing demand for food and increased use of pesticides and fertilisers, is resulting in soil depletion and erosion (Erisman et al., 2016). Unsustainable farming practices are not only threatening our biodiversity but also ecosystem services, upon which agriculture, our food-security itself, depends (Erisman et al., 2016). If these practices are not revolutionised, we are at significant risk of losing the vital ecosystems that we rely on for survival. The State of Nature Report identifies that the cumulative effect of fewer crops being planted, the loss of meadows, ponds and hedgerows, and a lack of crop rotation systems associated with mixed farming, has resulted in important wildlife habitats being negatively affected and the decline of many vital pollinators (Sustainable Food Trust, 2023).

Until recently there has been no legal requirement for farmers to maintain or improve biodiversity on their farms, although the use of pesticides, fertilisers and farm waste is monitored to reduce pollution and protect air quality in the UK. However, it is encouraging that the new Biodiversity Net Gain (BNG) strategy, which forms part of the Environment Act 2021(Legislation, 2021), will enable farmers to operate more sustainably. Moreover, farmers must no longer see conservation and farming as separate but rather incorporate changes across the whole farm (Sustainable Food Trust, 2023). An example of how one village is working with local farmers to explore some of these issues will be discussed in Chapter 12. It is imperative that changes in farming methods are implemented to encourage wildlife and halt the ecosystem crisis, without impacting on the growing demands for food production. Recognised internationally, sustainability in agriculture is one of the 17 Sustainable Development Goals developed by the UN (2015). So, how might these issues be addressed?

There are ways that farming can strike a balance with nature, and the more biodiverse a farm can be the healthier the system will be. The Soil Association (2023) believes that farms are more resilient to pest and disease outbreaks if they coexist with nature, and they support two principal approaches to biodiverse-enhancing farming. The first is a wildlife-friendly farming approach, where agricultural practices prioritise wildlife diversity, which enables a more integrated system. The second is land-sparing which is where agricultural land is split into different parcels; some are managed intensively for optimum yields, and others are allowed to return to a managed natural state. Other suggestions for farms and gardens to become hubs of diversity include: planting trees; sowing native wildflowers which encourage pollination and provide year-round food for wildlife; planting flowers and shrubs in different height groups which attract new year-round wildlife habitats who will set up home near their new food source, increasing overall biodiversity; creating more ponds which provide homes for dragonflies, damselflies and toads who are pest-control heroes (Soil Association, 2023).

Ocean pollution

Ocean or marine pollution is defined as chemicals and non-biodegradable rubbish that gets washed, spilled, pumped, leaked or blown into the oceans from the land, taking hundreds of years to decompose (National Geographic, 2023). Most chemical pollution results from fertiliser runoffs from farms (containing mainly nitrogen and phosphorus) which enters waterways and eventually flows into the ocean. These chemicals encourage the growth of algal blooms which are toxic to wildlife and harmful to humans, affecting food chains (National Geographic, 2023). Rubbish which enters the ocean comprises mostly plastic manufactured products such as shopping bags, food wrappers and bottles which can impact countless marine species and habitats. For example, plastic waste has been found on many remote coastlines, beaches and dive-sites worldwide. Scientists suggest that more than half of the world's sea turtles, and every seabird, will have eaten plastic at least once in their lifetime and animals, mistaking plastic for real food and tiny marine plankton, have been found to eat microplastics. These animals are then eaten by larger animals which absorb the plastic and toxic chemicals into their own tissues, and this eventually works its way up the food chain (Oceanic Society, 2023; National Geographic, 2023). Indeed, recent research has confirmed the presence of microplastics in human blood for the first time, including babies and young children who are more vulnerable to such exposure (Leslie et al., 2022). Separate research has also identified the presence of microplastics in human breast milk (Carrington, 2022) and semen (Carrington, 2024). With most packaging being made from plastic and ineffective waste management, the Oceanic Society (2023) estimates that between 4 and 12 million metric tons of plastic enters our oceans annually; enough to cover every metre of coastline in the world. If current waste production and management procedures remain unaltered, it is predicted that this annual amount could increase to between 12 and 36 million metric tons of plastic waste in the next 20 years (Oceanic Society, 2023).

Opportunities for action

There are many ways that we can reduce and even prevent marine pollution and several countries have introduced a zero-plastic policy. For example, Rwanda was the first country to become plastic-free and those found in possession of any plastic items could face criminal prosecution (Campbell, 2024). The Oceanic Society (2023) have compiled strategies that we can start to adopt today to make our oceans a safer and healthier place for all wildlife and ultimately, ourselves. These include rethinking what we eat and reducing our use of disposable or single-use plastic such as shopping bags, plastic bottles, cups and straws which will help keep plastics out of the ocean and reduce the quantity of new plastic in circulation.

Other opportunities include:

- Organising a beach, river, or local community clean-up as any plastic removed from the environment will reduce the amount of marine pollution. Do this individually, with friends and family or join a local group.
- Make a note of the non-recyclable waste you generate and work to source alternatives that are recyclable; visit your local recycling centre to find out what else can be recycled that you did not already know about. For those who live in the UK, 'The Big Plastic Count' (2024), takes place in March each year. This is an activity organised by Greenpeace and Everyday Plastic, which encourages households to count all plastic waste for a week, raising awareness of the amount produced. This is used as evidence to lobby government and press for stronger government commitments to reducing the use of plastic.
- Some international and national initiatives include The UK Plastics Pact (WRAP, n.d.) which is a collaboration of businesses working with the UK government and Non-Governmental Organisations (NGOs) to address the challenges of plastic waste through the development of a circular economy.

The impact of anthropogenic (human-made) climate change

Analysis of global climate records shows that the Earth's surface temperature has been rising since the beginning of the Industrial Period. Indeed, each decade since the 1980s has become increasingly warmer than the previous, with this trend expected to continue. The World Meteorological Office (WMO, 2023a) states "The global average near-surface temperature in 2023 was 1.45°C (with a margin of uncertainty of ±0.12°C) above the pre-industrial baseline." It is likely temperatures will temporarily exceed 1.5°C in the next five years (WMO, 2024). In addition, there were record-breaking sea surface temperatures in the North Atlantic and global ocean (The Copernicus Program, 2023). The most significant reason for this is the burning of fossil fuels. The resulting greenhouse gas emissions envelop the Earth, causing the sun's heat to become trapped and consequently raising temperatures. These greenhouse gases are primarily carbon dioxide (CO_2) and methane (CH_4) and they are the product of a range of human activities: Animal agriculture, (discussed further in Chapter 7), waste management, and fossil fuel production. Another serious effect of atmospheric pollution is Ocean Acidification, where seawater absorbs additional airborne CO_2 resulting in a reduced pH level of seawater which affects many marine organisms (NOAA, 2023). It reduces carbonate availability, having an osteoporosis-like effect on the shells and skeletons of calcium carbonate based marine life, reducing an organism's capacity to build and maintain themselves.

From a social justice perspective, countries that have contributed the least to global heating are already the most detrimentally affected. Simply put, the livelihood and health of people in low-to-middle-income countries are affected to a greater degree than those in high-income countries. The World Bank's Groundswell Report (Migration Data Portal, 2021) predicts that up to 216 million people will be forced to migrate within their countries by 2050 unless urgent steps are taken to reduce emissions. This includes Sub-Saharan Africa, North Africa, South Asia, Latin America, the Pacific, Central and Eastern Asia and Eastern Europe. Meanwhile, the UN (2023) warned that the world's largest fossil fuel producers (China, the USA, India, Russia and Japan), who already produce 42% of global emissions, are planning expansions that would exceed the planet's carbon budget by 200% (Carrington, 2023).

The United Nations Environment Programme (UNEP) Emissions Gap Report (2023a) underlines the seriousness of the current situation by emphasising the inadequacy of current carbon-cutting policies and, without serious commitment and interventions, an expected increase of 3°C of heating by the end of the century. The report highlights that existing 'net zero' pledges are not currently considered as credible, and that none of the G20 countries is currently on course to reduce the size of their emissions by the amount needed to achieve identified net zero targets by 2050. This is particularly concerning when G20 countries contribute 80% of CO_2 emissions. UNEP (2023b) writes,

> It is not only temperature increases that show the Earth's climate is getting warmer. Other climate changes occurring include melting polar ice and

glaciers, rising sea levels, warming oceans, changes in ocean chemistry and increases in extreme weather events.

Each of these are considered in further detail below.

Melting polar ice and glaciers

The global heating of recent decades has resulted in the glaciers and ice sheets at each pole starting to melt and shrink. The arctic permafrost has been warming and thawing since the 1970s, and glaciers have lost over 6,000 giga-tonnes in mass in the last few decades, while current analysis of data has shown, irrefutably, that Antarctic Sea ice fell to its lowest annual extent in February 2024 (WMO, 2023a; Earth Observatory, 2024). The consequences of this are discussed in more detail in Chapter 3.

Rising sea levels

Rising sea levels are the result of the increase in volume due to melting ice and glaciers, as well as water expansion following rising temperatures. This is reinforced by the Intergovernmental Panel on Climate Change (IPCC, 2021a) which has measured a 20cm rise in global sea levels in the last 100 years, 9cm of this during the last 30 years (Statista, 2023) and will continue to increase in line with greenhouse gas emissions. This is of particular concern for populations living within 10m above sea level. For example, countries such as the Maldives where even a 1-metre rise in sea levels will submerge over 80% of the remaining land (WHO, 2023). This is explored further in Chapter 4.

The UK is not immune to the impact of rising sea levels, which have risen approximately 6.5cm between 1981 and 2000 and are currently estimated to be rising at around 2.5 cm each decade (Climate Change Committee, 2021). The House of Lords Library (2021) states that approximately 5.2 million properties in England are currently at risk of flooding, and this could be set to double in the next 50 years if planning policies are not reviewed. It is estimated that coastal erosion will cause the loss of over 700 properties in England over the next 20 years, and 2,000 within 50 years; dependent upon existing sea defences being maintained to current standards (Surminski et al., 2016). However, if the existing defences were abandoned, around 500 properties would be lost in North Norfolk alone, together with the loss of some communities, resulting in the need to retreat inland due to coastal inundation or managed/non-managed coastal realignment (Arnall, 2023; Dawson et al., 2009; Koukoulas et al., 2005). Already there are casualties as Gwynedd Council confirmed that it was no longer able to maintain flood defences for the village of Fairbourne, which is expected to become uninhabitable by 2054 as it returns to marshland (Gerretsen, 2022).

Globally, coastal flooding will see increased numbers of people displaced from their homes and trade and supply chains affected. The challenges for displaced people because of climate change will be explored further in Chapter 4: Climate migration and displacement.

Warming oceans

Ocean temperatures are rising. Recent studies show that upper ocean warming accounts for an estimated 63% increase in the total amount of stored heat in the climate system from 1971 to 2010, and deeper warming from 700m to the ocean floor adds a further 30% (Climate. Gov, 2023). The latest analysis shared by WMO (2023b) states that the upper 2000m of the ocean is continuing to warm in response to this, and The Copernicus Program (2024) reports global ocean temperature in 2024 was the highest on record.

Changes in ocean chemistry

The oceans are a storage tank for heat and also a 'carbon sink.' In other words, they absorb CO_2 from the atmosphere; approximately 30% of the total carbon emissions (CSIRO, 2023). Consequently, the last 40 years have seen increases in both the total amount of carbon emissions and global ocean heat content (OHC). The challenge with this particular 'carbon sink' is that due to increased CO_2 absorption, oceans are becoming more acidic, at a rate ten times faster than at any other point in the last 65 million years (IPCC, 2021b). This is because CO_2 and water combine to form carbonic acid (H2CO3), which is a weak acid that breaks (or "dissociates") into hydrogen ions (H+) and bicarbonate ions (HCO3–) (NOAA, 2023). The resulting chemical reaction impacts many marine species, particularly organisms with hard shells and skeletons (made of calcium and carbonate from seawater), such as oysters, corals and pteropods. As ocean acidification continues this results in fewer carbonate ions available for the building and maintaining of these hard shells or skeletons of carbonate organisms. If acid levels continue to increase further, their shells and skeletons will continue to dissolve, and these organisms will be at risk of extinction. They are an important part of many food webs and if these disappear other species, from tiny krill to gigantic whales, will be put at risk. Furthermore, it is not only calcifying organisms which will be put at risk. For instance, larval clownfish, the basis of many food webs, will struggle to find a habitat in acidic waters and this will inevitably affect their numbers significantly (NOAA, 2023).

Increase in extreme weather events

Undisputably, cases of extreme weather events such as heatwaves and storms (including heavy rainfall and droughts), are increasing around the globe (WMO,

2023c) and will become more frequent, and more severe, as greenhouse gas emissions and global temperatures continue to rise (IPCC, 2021c). Vu (2024) suggests that the current categories of storms need to expand to include a sixth category "for storms with winds that exceed 192mph (309kmph)," acknowledging the growing intensity of tropical weather systems.

Examples of global extreme weather events in the past couple of years include: raining rather than snowing at the peak of the Greenland Ice Sheet in August 2021 for the first time on record (European Space Agency, 2022); a Canadian heatwave which saw temperatures top 49.6C in Lytton, British Columbia in July 2021 (British Broadcasting Corporation (BBC), 2021a); and in July 2021, a year's worth of rain fell on Zhengzhou in central China over just 3 days (BBC, 2021b); Storm Ana in January 2022 which washed away roads in Madagascar, Mozambique and Malawi, cutting off thousands of people; flooding in India from June to September 2022 which saw floodwater and water-borne diseases affect a third of the country and impacted more than 33 million people (BBC, 2022); the summer of 2023 saw severe heat waves in parts of the US, Europe and Asia, with areas Sicily and Sardinia in Italy reaching nearly 48C; parts of Greece experienced severe drought followed by wildfires causing fatalities, casualties and economic loss (BBC, 2023); in South America, floods in Brazil (BBC, 2024) and wildfires in Chile (Al-Jazeera, 2024) have killed many and displaced thousands.

Even if current greenhouse gas emissions are maintained at existing rates, sea levels will continue to rise beyond 2100 and have an impact for centuries (IPCC, 2019). Meanwhile, if global heating continues it could lead to the complete loss of the Greenland Ice sheet which would cause global mean sea levels to rise by up to seven metres (IPCC, 2013). Carrington et al. (2023) have produced a useful resource of ten charts which illustrate many of the changes that have been discussed.

Opportunities

Energy

Thankfully, there are strategies and actions that can be taken. Risks can be reduced by accurate forecasting, increasing property-level protection and sea defences, improved land-use planning, and managed coastline realignment, as discussed in Chapter 11. Additionally, individual, organisational and community-led projects focused on improving their own protection levels support the empowerment of CaYP in their communities. An example of how one small community group is working to address some of these issues is explored in Chapter 12. The UK, due to its long coastlines, shallow waters, and strong winds, has the greatest number of off-shore wind turbines in the world. As of 2023 the UK had a total of 11,000 wind turbines, both on- and off-shore

(Ukpanah, 2024). One turbine revolution has the potential to generate sufficient energy to power a house for a day. Already, 39.4% of UK electricity is generated from renewable resources (Maguire, 2024) and electricity generated from wind power has increased 715% between 2009 and 2020 (ONS, 2021). Other countries involved in positive action include Denmark which is generating 40% of energy through wind turbines (Clean River Recycling Solutions, 2024).

Food

Positive international action to mitigate global heating may help to slow and potentially reduce the overall rate of sea level rise and sea level extremes in the long term. This includes the development and use of renewable energy and encouraging populations to change practices and behaviours, such as a move towards more sustainable diets. As consumers we can all take positive steps when trying to develop a more sustainable diet including, aim to eat everything we buy, avoid air-freighted food, reduce meat and dairy consumption – especially lamb and beef; all of which will help to reduce the environmental impact. This will be explored in Chapter 7.

Conclusion

This chapter has explored what we mean when referring to climate change and biodiversity, the changes that have occurred since the Industrial Revolution and the impact of this on our world, both nationally and globally. The data is alarming, but there are elements of our future that remain unwritten. We can still make changes, and we have started to explore some of these possibilities in this chapter. As the Global Climate Change Panel (2024) states,

> While the effects of human activities on Earth's climate to date are irreversible on the timescale of humans alive today, every little bit of avoided future temperature increase results in less warming that would otherwise persist for essentially forever.

Meanwhile, we can all take that first step. Find out your own carbon footprint using the WWF calculator "How Big Is Your Environmental Footprint?" and identify what you can do to make a difference. As Barack Obama (2014) reminds us, "We are the first generation to feel the impact of climate change and the last generation that can do something about it."

References

Arnall, A (2023) Encountering the anthropocene: Reconfiguring human-nature relations on the North Norfolk Coast, UK, Geoforum. *Science Direct* 143, 103768. https://doi.org/10.1016/j.geoforum.2023.103768.

British Broadcasting Corporation (2021a) *Canada Lytton: Heatwave record village overwhelmingly burned in wildfire.* Available at: https://www.bbc.co.uk/news/world-us-canada-57678054 (Accessed 12 February 2024).

British Broadcasting Corporation (2021b) *China floods: 12 dead in Zhengzhou train and thousands evacuated in Henan.* Available at: https://www.bbc.co.uk/news/world-asia-china-57861067 (Accessed 12 February 2024).

British Broadcasting Corporation (2022) *Storm Ana: Deadly Africa storm shows climate crisis reality – UN.* Available at: https://www.bbc.co.uk/news/world-africa-60180738 (Accessed 12 February 2024).

British Broadcasting Corporation (2023) *The extreme summer weather that scorched and soaked the world.* https://www.bbc.co.uk/news/resources/idt-8f0357f9-9013-4567-8407-be938c8c70cf (Accessed 12 February 2024).

British Broadcasting Corporation (2024) *Brazil floods: Hundreds of Grande do Sol towns under water.* Available at: Brazil floods: Hundreds of Rio Grande do Sul towns under water. Available at: https://www.bbc.co.uk/news/world-latin-america-68968987#:~:text=Heavy%20rains%20which%20caused%20widespread,from%20their%20homes%2C%20officials%20said. (Accessed 22 February 2025).

British Geological Survey (2023) *Discovering geology.* Available at: https://www.bgs.ac.uk/discovering-geology/ (Accessed 10 July 2023).

Campbell, L. (2024) *Top 10 countries doing the most to tackle plastic pollution.* Available at: https://good-with-money.com/2024/01/29/top-10-countries-doing-the-most-to-tackle-plastic-pollution/# (Accessed 6 February 2024).

Carrington, D. (2022) *Microplastics found in human breast milk for the first time.* Available at: https://www.theguardian.com/environment/2022/oct/07/microplastics-human-breast-milk-first-time (Accessed 20 October 2024).

Carrington, D. (2023) *World facing 'hellish' 3C of climate heating, UN warns before Cop28.* Available at: https://www.theguardian.com/environment/2023/nov/20/world-facing-hellish-3c-of-climate-heating-un-warns-before-cop28#:~:text=The%20UN%20warned%20earlier%20in,which%20experts%20called%20%E2%80%9Cinsanity%E2%80%9D (Accessed 23 January 2024).

Carrington, D. (2024) *Microplastics found in every human semen sample tested in study.* Available at: https://www.theguardian.com/environment/article/2024/jun/10/microplastics-found-in-every-human-semen-sample-tested-in-chinese-study#:~:text=Microplastics%20found%20in%20every%20human%20semen%20sample%20tested%20in%20study,-This%20article%20is&text. Available at: =Microp (Accessed 20 October 2024).

Carrington, D., Leach, A, Scruton, P. and Symons, H. (2023) *The climate crisis explained in 10 charts.* Available at: https://www.theguardian.com/environment/2023/nov/30/the-climate-crisis-explained-in-10-charts-co2-green-energy-cop28 (Accessed 13 December 2023).

Clean River Recycling Solutions (2024) *6 countries leading the way in sustainability.* Available at: https://cleanriver.com/resource/blogleading-countries-in-sustainability/#:~:text=Denmark%20announced%20they%20have%20set,including%20solar%2C%20wind%20and%20biomass (Accessed 9 July 2024).

Climate Change Committee (CCC) (2021) *Independent assessment of UK climate risk.* Available at: https://www.theccc.org.uk/wp-content/uploads/2021/07/Independent-Assessment-of-UK-Climate-Risk-Advice-to-Govt-for-CCRA3-CCC.pdf (Accessed 8 September 2023).

Copernicus Program (2023) *Global sea-surface temperatures reach record high*. Available at: https://climate.copernicus.eu/global-sea-surface-temperature-reaches-record-high (Accessed 30 October 2023).

Copernicus Program (2024) *Copernicus: May 2024, streak of global records for surface air and ocean temperatures continues*. Available at: https://climate.copernicus.eu/copernicus-may-2024-streak-global-records-surface-air-and-ocean-temperatures-continues (Accessed 12 October 2024).

CSIRO (2023) *Oceans absorb 30% of our emissions, driven by a huge caron pump*. Available at: https://www.csiro.au/en/news/all/articles/2023/june/oceans-absorb-emissions (Accessed 4 February 2024).

Dawson, R.J., Dickson, M.E., Nicholls, R.J., Hall, J.W. et al. (2009) Integrated analysis of risks of coastal flooding and cliff erosion under scenarios of long-term change. *Climatic Change* 95(1–2), 249–288. https://doi.org/10.1007/s10584-008-9532-8.

Department for Environment Food and Rural Affairs (2023) *Agricultural land use in United Kingdom at 1 June 2023*. Available at: https://www.gov.uk/government/statistics/agricultural-land-use-in-the-united-kingdom/agricultural-land-use-in-united-kingdom-at-1-june-2023#:~:text=The%20utilised%20agricultural%20area%20(UAA,third%20(36%25)%20of%20UAA (Accessed 12 October 2024).

Earth Observatory (2024) *Atlantic sea-ice at near-historic lows*. Available at: Antarctic Sea Ice at Near-Historic Lows (nasa.gov) (Accessed 27 September 2024).

Erisman, J.W., van Eekeren, N., De Wit, J., Koopmans, C.J., Cuijpers, W.J.M., Oerlemans, N. and Koks, B.J. (2016) Agriculture and biodiversity: A better balance benefits both. *AIMS Agriculture and Food* 1, 157–174. https://doi.org/10.3934/agrfood.2016.2.157.

European Space Agency (2022) *Historic Greenland ice sheet rainfall unravelled*. Available at: https://www.esa.int/Applications/Observing_the_Earth/FutureEO/Historic_Greenland_ice_sheet_rainfall_unravelled (Accessed 12 February 2024).

Gerretsen, I. (2022) *The UK 'climate refugees' who won't leave*. Available at: https://www.bbc.com/future/article/20220506-the-uk-climate-refugees-who-wont-leave (Accessed 6 February 2024).

Global Climate Change Panel (2024) *Is it too late to prevent climate change?* Available at: https://climate.nasa.gov/faq/16/is-it-too-late-to-prevent-climate-change/#:~:text=While%20the%20effects%20of%20human,otherwise%20persist%20for%20essentially%20forever (Accessed 9 February 2024).

House of Lords Library (2021) *House developments on functional flood plains*. Available at: https://lordslibrary.parliament.uk/housing-developments-on-functional-flood-plains/ (Accessed 4 February 2024).

Intergovernmental Panel on Climate Change IPCC (2013) *Climate change 2013 fifth assessment report of the intergovernmental panel on climate change wg i intergovernmental panel on climate change*. Available at: https://www.ipcc.ch/site/assets/uploads/2018/03/WG1AR5_SummaryVolume_FINAL.pdf. (Accessed 22 February 2025).

Intergovernmental Panel on Climate Change (IPCC) (2019) Summary for policymakers. *IPCC special report on the ocean and cryosphere in a changing climate*. Available at: https://www.ipcc.ch/site/assets/uploads/sites/3/2022/03/01_SROCC_SPM_FINAL.pdf (Accessed 4 February 2024).

Intergovernmental Panel on Climate Change (IPCC) (2021a) *Climate change 2021: The physical science basis, technical summary, box TS.4*. Available at: https://www.ipcc.ch/report/ar6/wg1/chapter/summary-for-policymakers/ (Accessed 18 July 2023).

Intergovernmental Panel on Climate Change (IPCC) (2021b) *Climate change 2021: The physical science basis, Chapter 2, Section 2.3.3.5*. Available at: https://www.ipcc.ch/report/ar6/wg1/downloads/report/IPCC_AR6_WGI_FullReport_small.pdf (Accessed 22 February 2025).

Intergovernmental Panel on Climate Change (IPCC) (2021c) *Climate change 2021: The physical science basis, Section A.3 and B.2*. Available at: https://www.ipcc.ch/report/ar6/wg1/downloads/report/IPCC_AR6_WGI_FullReport_small.pdf (Accessed 22 February 2025).

Koukoulas, S., Nicholls, R.J., Dickson, M.E., Walkden, M. et al. (2005) A GIS tool for analysis and interpretation of coastal erosion model outputs (SCAPEGIS). In *Coastal Dynamics 2005. 5th International Conference on Coastal Dynamics*. Virginia, USA: American Society of Civil Engineers.

Legislation (2021) *Environment act*. Available at: Environment Act 2021 (legislation.gov.uk) (Accessed 16 October 2024).

Leslie, H.A., van Velzen, M.J.M., Brandsma, S.H., Vethaak, A.D., Garcia-Vallejo, J.J., Lamoree, M.H. (2022) Discovery and quantification of plastic particle pollution in human blood. In *Environment International*, 163. Available at: https://www.sciencedirect.com/science/article/pii/S0160412022001258 (Accessed 6 February

Maguire, G. (2024) *Wind overtakes fossil fuels for UK electricity generation*. Available at: https://www.reuters.com/business/energy/wind-overtakes-fossil-fuels-uk-electricity-generation-maguire-2024-04-23/ (Accessed 20 October 2024).

Migration Data Portal (2021) *Environmental migration*. Available at: https://www.worldbank.org/en/news/press-release/2021/09/13/climate-change-could-force-216-million-people-to-migrate-within-their-own-countries-by-2050 (Accessed 6 February 2024).

National Biodiversity Network (2019) *State of nature 2019 UK full report*. Available at: https://nbn.org.uk/wp-content/uploads/2019/09/State-of-Nature-2019-UK-full-report.pdf (Accessed 2 November 2023).

National Geographic (2022) *Earth's changing climate*. Available at: (Accessed 10 July 2023).

National Geographic (2023) *Marine pollution*. Available at: https://education.nationalgeographic.org/resource/marine-pollution/ (Accessed 2 November 2023).

National Oceanic and Atmospheric Administration (NOAA), (2023). *Ocean acidification*. Available at: https://www.noaa.gov/education/resource-collections/ocean-coasts/ocean-acidification (Accessed 18 July 2023).

Natural England (2023) *Natural England unveils New Green Infrastructure Framework*. Available at: https://www.gov.uk/government/news/natural-england-unveils-new-green-infrastructure-framework (Accessed 12 October 2024).

Natural History Museum (2022) *Destruction of forests and grasslands is biggest cause of biodiversity loss*. Available at: https://www.nhm.ac.uk/discover/news/2022/november/destruction-forests-and-grasslands-biggest-cause-of-biodiversity-loss.html (Accessed 12 February 2024).

Natural History Museum (2023) *What is biodiversity?* Available at: https://www.nhm.ac.uk/discover/what-is-biodiversity.html (Accessed 1 November 2023).

NatureScot (2023). *What are ecosystem services?* Available at: https://www.nature.scot/scotlands-biodiversity/scottish-biodiversity-strategy-and-cop15/ecosystem-approach/ecosystem-services-natures-benefits#:~:text=Ecosystem%20Services%20are%20the%20direct,as%20reducing%20stress%20and%20anxiety (Accessed 8 February 2024).

NOAA (2022) *Marine pollution: From and through the atmosphere.* Available at: https://www.noaa.gov/gc-international-section/marine-pollution-from-and-through-atmosphere#:~:text=One%20serious%20consequence%20of%20atmospheric,may%20affect%20many%20marine%20organisms (Accessed 1 February 2024).

Obama, Barack (2014) *Remarks by the President at U.N. Climate change summit.* Available at: https://obamawhitehouse.archives.gov/the-press-office/2014/09/23/remarks-president-un-climate-change-summit (Accessed 5 June 202).

Oceanic Society (2023) *7 ways to reduce ocean plastic.* Available at: https://www.oceanicsociety.org/resources/7-ways-to-reduce-ocean-plastic-pollution-today/ (Accessed 2 November 2023).

Office for National Statistics (2022) *Excess mortality during heat periods.* Available at: https://www.ons.gov.uk/peoplepopulationandcommunity/birthsdeathsandmarriages/deaths/articles/excessmortalityduringheatperiods/englandandwales-1juneto31august2022/previous/v1#excess-mortality-during-heat-periods-data (Accessed 8 September 2023).

ONS (2021) *Wind energy in the UK: 2021.* Available at: https://www.ons.gov.uk/economy/environmentalaccounts/articles/windenergyintheuk/june2021 (Accessed 9 February 2024).

Soil Association (2023) *Can agroecology increased biodiversity on farms.* Available at: https://www.soilassociation.org/blogs/2021/may/19/can-agroecology-increase-biodiversity-on-farms/ (Accessed 2 November 2023).

Statista (2023) *Mean sea level variation worldwide from 1993 to 2023.* Available at: https://www.statista.com/statistics/603821/global-cumulative-sea-level-rise/#:~:text=In%20January%202023%2C%20mean%20sea,melting%20ice%20caps%20and%20glaciers (Accessed 1 February 2024).

Surminski, S., Baglee, A., Cameron, C., Connell, A. et al. (2016) *Chapter 6, business and industry.* In *UK climate change risk assessment evidence report.* Prepared for the Adaptation Sub-Committee of the Committee on Climate Change. London.

Sustainable Food Trust (2023) *Bringing biodiversity back into farming.* Available at: https://sustainablefoodtrust.org/news-views/bringing-biodiversity-back-into-farming/ (Accessed 2 November 2023).

The Big Plastic Count (2024) *Let's investigate to stop the climate crisis.* Available at: https://thebigplasticcount.com/ (Accessed 12 February 2024).

Tobi, R., Driscoll, M., Gurung, I. and English, A. (2023) *Low income, low emissions.* Available at: https://foodfoundation.org.uk/sites/default/files/2023-07/TFF_CLIMATE%20BRIEFING.pdf (Accessed 9 February 2024).

Ukpanah, I. (2024) *The statistical landscape of wind farms in the UK.* Available at: https://www.greenmatch.co.uk/green-energy/wind-farm (Accessed 20 October 2024).

UN (2015) *Sustainable consumption and production.* Available at: https://sdgs.un.org/topics/sustainable-consumption-and-production (Accessed 2 November 2023).

United Nations (2023) *What is climate change?* Available at: https://www.un.org/en/climatechange/what-is-climate-change (Accessed 10 July 2023).

United Nations Environment Programme (UNEP) (2023a) *Emissions gap report. Broken record temperatures hit new highs, yet world fails to cut emissions (again).* Available at: https://www.unep.org/resources/emissions-gap-report (Accessed 21 November 2023).

United Nations Environment Programme (2023b) *Nations must go further than current Paris pledges or face global warming of 2.5-2.9'C*. Available at: https://www.unep.org/news-and-stories/press-release/nations-must-go-further-current-paris-pledges-or-face-global-warming (Accessed 18 June 2024).

Vu (2024) *In a warming world, climate scientists consider category 6 hurricanes*. Available at: https://crd.lbl.gov/news-and-publications/news/2024/in-a-warming-world-climate-scientists-consider-category-6-hurricanes/ (Accessed 12 October 2024).

WHO (2023) *Maldives: Leading the way to sustainable environmental health*. Available at: https://www.who.int/southeastasia/news/feature-stories/detail/sustainable-environmental-health-maldives (Accessed 6 February 2024).

World Meteorological Organization (WMO) (2023a) *WMO global annual to decadal climate update*. Available at: https://hadleyserver.metoffice.gov.uk/wmolc/WMO_GADCU_2023-2027.pdf (Accessed 17 July 2023).

World Meteorological Organization (WMO) (2023b) *State of the climate 2022*. Chapter: 'Cryosphere', pp. 13–19. Available at: https://www.who.int/southeastasia/news/feature-stories/detail/sustainable-environmental-health-maldives (Accessed 23 February 2025)

World Meteorological Organization (2023c) *2023 shatters climate records with major impacts*. Available at: https://wmo.int/news/media-centre/2023-shatters-climate-records-major-impacts (Accessed 9 July 2024).

World Meteorological Organization (WMO) (2024) *Global temperature is likely to exceed 1.5°C above pre-industrial level temporarily in next 5 years*. Available at: https://wmo.int/news/media-centre/global-temperature-likely-exceed-15degc-above-pre-industrial-level-temporarily-next-5-years (Accessed 27 September 2024).

WWF (n.d.) *How big is your environmental footprint?* Available at: https://footprint.wwf.org.uk/ (Accessed 9 February 2024).

PART 2
Pedagogies of hope

Nicola Watson

In this section, multiple perspectives are explored on how those who work with children and young people might respond to the environmental crisis through their work. Each chapter presents a unique response to a specific focus. We owe it to children and young people to acknowledge the reality of the environmental crisis and the validity of their emotional responses. Hence, the section begins with a chapter on nurturing and supporting good mental health in a time of uncertainty. What unites the chapters in this section is a recognition of the overriding necessity of hope. Hope is strengthened by the pedagogies which demonstrate commitment to empowerment and agency, inclusion and social justice, the four themes threaded throughout the book.

Pedagogies which value the unique ideas and positive contributions of individuals, however small, nurture the conditions for empowerment. To be able to make changes and act constructively in response to challenges is tremendously powerful in supporting good mental health and conveying the message to children and young people that they are intrinsically valuable and valued.

Looking beyond the individual, the concept of interconnection is explored. It invites us to widen our ideas on the issue of inclusion and to resist the construct that personal and national self-interests can be regarded in isolation. Radical inclusion involves a recognition that the interests of vulnerable groups such as those displaced by climate change and the more-than-human whose habitats are being destroyed, are inextricable from the interests of all on a fragile planet with finite resources. Making connections between our shared interests and our interdependence, provides us with rich resources for positive action, individually, in community and globally.

DOI: 10.4324/9781003506850-4

The case studies included in this section are drawn intentionally, from a wide range of populations and locations. They are aimed at highlighting commonalities rather than differences. They demonstrate that sustainable practice can be seeded anywhere and vignettes from cultures around the world can be reflected upon, interpreted and applied in your own contexts where you can help them to flourish.

2

ECO-ANXIETY

A positive response

Alison Prowle and Lowri Prowle

Introduction

We live in frightening times. With global warming, diminishing natural resources, rising sea levels and melting icecaps, it is natural for individuals to experience elevated levels of anxiety. Hickman et al. (2021) view climate anxiety not as psychological dysfunction, but rather as an entirely rational and compassionate response to a very real and serious problem. This may be compounded by frustration and anger that others (including politicians and powerful economic forces) do not share their concerns, or at the very least, do not recognise the urgency. However, if not sensitively addressed and supported, the concerns of children and young people (CaYP), about the environmental crises could cause significant mental health problems that may become overwhelming and all consuming (Clayton and Karazsia, 2020). Hence this chapter is in two parts. The first section explores eco-anxiety and how it may manifest itself in children's lives. Here we highlight helpful interventions to support children's wellbeing and help them to manage difficult emotions. The second section considers how engagement, participation and activism may support children and young people to move beyond feelings of fear and hopelessness and find positive ways to contribute to the climate debate and climate action.

What is eco-anxiety?

Eco-anxiety refers to an individual's persistent worry about the future of the earth and the life it shelters (Coffey et al., 2021). Although not yet considered a diagnosable condition within the Diagnostic Manual of Mental Disorders

DOI: 10.4324/9781003506850-5

(American Psychiatric Association, 2022), recognition of eco-anxiety and its complex psychological effects is rapidly growing (Kurth and Pihkala, 2022). While typical anxiety disorders are often characterised by worrying about hypothetical scenarios (events that will probably never occur), eco-anxiety is unique in the sense that worrying about the future of the earth is a legitimate and understandable response considering the current climate emergency. However, even within this context, it might be possible to categorise worries into those that can be solved and those that cannot be solved. For young people, a way of dealing with this anxiety might involve becoming active in the areas where their actions can make a positive difference.

It is understood that eco-anxiety disproportionately impacts children and young people (Léger-Goodes et al., 2022). As we see in the case study later in this chapter, parents and educators are becoming more aware of how children are responding to the unfolding climate situation. To be diagnosed with any anxiety disorder a strong feeling of fear, worry or unease that persists and interferes with day-to-day life must be evident (American Psychiatric Association, 2022). Climate anxiety, specifically, can manifest itself in the following ways: anxiety about personal impact on the planet and constant worrying about the future of the earth (Verplanken et al., 2020) and/or continual monitoring of news outlets (Coffey et al., 2021). Eco-anxiety can also manifest physically through symptoms such as stomach aches, headaches, muscle pains and difficulty sleeping (Clayton and Karazsia, 2020). Much of how CaYP express eco-anxiety will vary based on factors including the age of the child, the social context, culture and exposure to ecological issues. Beyond the immediate impacts there may be further implications of eco-anxiety including negative impacts on friendships, school performance, abilities to pursue hobbies and interests. The vignettes below (constructed from our combined professional experience) provide some insights into how climate anxiety might present in children of different ages.

Case study: Bobby (aged 8)

Bobby lives with his mum and sister in Monmouth. He is a very curious and enthusiastic child who loves learning about new things. Bobby sometimes watches the news with his mum and sister after school. As a result, he has become increasingly aware of environmental issues and with every news report about melting ice caps, or forests disappearing, he has become more anxious. On a day out at the local agricultural show, Bobby also found out that his local river is polluted and the wildlife in the river has been affected. Recently Bobby has been struggling to sleep at night and as a result is struggling to concentrate at school. Another parent has mentioned to the teacher that her daughter came home from school in tears because Bobby told her the world was going to end.

Case study: Rina (aged 15)

Rina lives with her family in South London. She performed well in her early GCSEs and is captain of the girls' netball team. However, recently teachers have noticed that Rina's grades are slipping, she has withdrawn from her extra-curricular activities and has become more isolated from her friendship group. Rina has spoken to the school counsellor, and it became known that she is very anxious about climate change. She reports spending her evenings watching TikTok videos about the climate emergency. Rina has relatives in Bangladesh, an area heavily affected by rising sea levels. There have been many family discussions about the safety of family members and their livelihoods. Rina states that she has lost interest in schoolwork because she does not see the point when the future is so uncertain.

Supporting children with eco- anxiety

As is demonstrated above, worry about the future of the planet can affect anyone and is a legitimate concern. We do not want to pathologise children who take an active interest in the future of the planet, which determines their future. However, we recognise that worries can become overwhelming and unmanageable for children and young people. Below we suggest interventions that may be helpful when a child's eco-anxiety begins to have a profound impact on their day-to-day life. Many of the interventions suggested to treat eco-anxiety will be akin to those used to treat more generalised anxiety disorders. We explain what these interventions are and how they can be implemented by parents and educators When implementing these interventions, it is important to consider the age and development of the child.

Mindfulness

Mindfulness is a technique which involves paying full attention to your thoughts, feelings, and bodily sensations (Creswell, 2017). When it comes to anxiety, mindfulness aims to bring attention back to the present moment to minimise negative rumination (Hofmann and Gómez, 2017). Mindfulness is one of the fastest growing interventions in adult mental health (Prakash et al., 2017) but more recently there has been a focus on the effectiveness of mindfulness for CaYP (Saunders and Kober, 2020). Research supports the use of mindfulness-based interventions in the treatment of anxiety for children as young as five (Ruiz-Íñiguez et al., 2020). There is also an emerging evidence base which supports the use of mindfulness as an intervention for eco-anxiety specifically (Baudon and Jachens, 2021). Parents and educators can support children to become mindful through a variety of different techniques and approaches. Some of these are outlined below:

- Breathing exercises such as counted breath or box breathing. Box breathing is a mindful relaxation technique which involves focusing on a square object and following a four-sided breathing pattern (inhale, hold, exhale, hold). This technique can be particularly helpful in easing panic and controlling hyperventilation (Ahmed et al., 2021)
- 5,4,3,2,1 method which involves identifying five things you can see, four things you can touch, three things you can hear, two things you can smell, and one thing you can taste
- Mindful consumption of food where the individual is encouraged to spend time focusing on a food item, taking account of its texture, taste, smell before eating it.

There is an extensive literature base that parents and teachers can draw upon to gain further information about using mindfulness techniques with children.

Cognitive Behavioural Therapy (CBT) approaches

Cognitive Behavioural Therapy (CBT) is based on the concept that your physical feelings, thoughts and behaviours are interlinked in a negative cycle, which can perpetuate emotional difficulties (Greimel and Kröner-Herwig, 2011). CBT approaches can provide the individual with techniques for managing difficult emotions such as fear or anger (Öst and Clark, 2013). A large body of research has shown that CBT is effective in treating childhood anxiety disorders, but in recent years, research has also supported the idea that CBT can be adapted and applied for children experiencing symptoms of eco-anxiety (James et al., 2020). There are a variety of CBT techniques that parents and educators can introduce when working with children who suffer from eco-anxiety. These include:

- Worry time: Setting a discrete period of time each day to focus entirely on worries and try to identify solutions. Children should then be encouraged to refocus attention during the day and save worrying until 'worry time' (Dippel et al., 2024).
- Psychoeducation: This is a core element of CBT (Dolan et al., 2021). The aim is to educate the child on common symptoms of anxiety and the function of this anxiety. Having this knowledge can help to normalise the experience of anxiety and empowers the child or young person to work towards developing positive coping strategies.

- Worry box: A worry box is a container where a child can safely store their worries. They can write or draw their anxious thoughts and place them into the box. This encourages the child to let go of the worry and focus on the present moment (Creswell et al., 2020).

Physical activity

Getting active is acknowledged as a crucial aspect of maintaining wellbeing (Thompson et al., 2011) and research suggests that exercise can serve as a protective factor against anxiety disorders in children and young people (Wegner et al., 2014). Physical activity can enhance wellbeing in a number of ways: moving the body can reduce muscle tension and other physical symptoms of anxiety (Kandola and Stubbs, 2020), it can alter brain chemistry, increasing the supply of anti-anxiety neuro-chemicals such as serotonin (Kondo and Shimada, 2015) and at the most basic level engaging in exercise can act as a useful distraction, diverting the mind away from worrying. Parents and educators should encourage CaYP to get active in ways that synergise with the individual child's hobbies and interests (Hodgkins and Prowle, 2023). Examples include those below, but the possibilities for being active are endless:

- Outdoor play
- Sports clubs
- Swimming
- Nature walks

Of course, for many children living in urban areas of deprivation, such opportunities may be limited, but there does seem to be an increased understanding of the importance of urban green spaces and an impetus among local planning agencies to preserve and create them in order to support community wellbeing (Wolch et al., 2014).

Maintaining positive relationships with others

As was seen in the second vignette, children experiencing eco-anxiety often find themselves withdrawing from others. However, as Hart et al. (2016) indicates within her resilience framework, positive relationships with peers and adults can provide a strong protective factor for children and young people experiencing mental health difficulties. There are of ways in which children can be supported in this area as we outline below:

- Intergenerational activities: Providing an opportunity to gain experience from and connect with the older generation. This could be spending time with grandparents or older adults within the wider community. Activities might include cooking, gardening or playing an instrument.

- Enhancing peer friendships: This could include attendance and clubs and after school activities, participation and group projects, and opportunities for play.
- Ensuring children within the education setting have access to a trusted adult, (this could be a teacher, teaching assistant, or a member of the pastoral care team) who can provide support and signposting.

Nature connection

Encouraging children to spend time outdoors and to connect with nature in a mindful way can be helpful in countering the worst effects of eco-anxiety. Living near nature or having green school yards is associated with less stress and higher psychological wellbeing among children (Chawla et al., 2014). Additionally, children and young adults who frequently spend time outdoors present with reduced anxiety symptoms compared to those who do not (Gladwell et al., 2013). Parents or educators might facilitate these opportunities through a number of ways such as gardening, visiting parks and natural places or looking at the sky at night. While this is easier for families living in rural communities, there will be opportunities for outdoor adventures even in urban landscapes.

Research has also found that spending time outdoors leads to an increase in children's desires to protect nature (Cheng and Monroe, 2012). Therefore, by fostering a deeper connection with the natural world children may become more empowered to make positive changes and less anxious about the state of the environment. We will revisit this later in this chapter when we explore the power of activism.

One of the simplest ways of summarising the ideas outlined above is through the NHS five ways to wellbeing model (NHS, 2022). The idea is that wellbeing can be improved and maintained through five core steps; connect with others, be physically active, learn new skills, give to others, pay attention to the present moment. This supports the importance of taking a holistic approach and building on different areas within an individual's life in order to yield positive outcomes for those who struggle with eco-anxiety.

Although this acts as a general guide to supporting children with eco-anxiety, it is important that parents and practitioners seek professional help if symptoms become unmanageable.

The suggestions that we outline above could be referred to as a 'clinical' response to supporting CaYP with anxiety, and this is important, particularly when anxiety is acute. However, in thinking about how we support CaYP with eco-anxiety more holistically, we will want to move beyond treating the symptoms, to enabling children to make a positive contribution and be empowered and able to face the future with hope. The following sections consider some of the ways that we can help CaYP understand the big questions that surround the climate emergency, manage uncomfortable emotions, and collaborate with others to make sustainable choices and changes.

Taking action

One of the most powerful ways that children can deal with their eco-anxiety is through getting involved in action that supports the environment. Environmental activism is defined as "Any form of activity by environmentalists that is designed to raise public awareness of environmental issues" (Oxford Reference, 2024). Before delving further into different forms that such activism might take, it is helpful to recognise that there are excellent examples of young people advocating for more sustainable ways of living.

Role models for young people

Children experiencing eco-anxiety can be inspired by the stories of other young people who actively work to raise awareness of environmental issues and campaign for change (Chauzy, 2020, n.p.). Notable examples of child eco-activists include:

- Greta Thunberg: Perhaps the most well-known child eco-activist, Greta Thunberg gained international recognition for her solo climate strike outside the Swedish parliament in 2018 when she was just 16 years old. Since then, she has inspired millions of people around the world to join the Fridays for Future movement, advocating for urgent climate action from governments and businesses across the globe.
- Autumn Peltier: Autumn is an activist for clean water and Indigenous communities. At 13 years of age she addressed the UN General Assembly, assertively making the point that "I've said it once, and I'll say it again we can't eat money or drink oil."
- Leah Namugerwa: Leah is a young Ugandan climate activist who is lobbying her government to act against the climate crisis by launching the campaign #BanPlasticUG to tackle the issue of plastic pollution. She was first made aware of the impact of climate change when watching local news about prolonged drought and landslides in Uganda.
- Lesein Mutunkei: Lesein is a 15-year-old from Kenya who has made his love of football into an incentive for environmental action. He plants a tree for every goal he scores. Lesein has encouraged his school and football club to be more sustainable, and he hopes to expand his #Trees4Goals campaign across Africa.

Often, the starting point for young people's activism is education or awareness raising through reading, news outlets or social media. Becoming cognisant of the issues can then provide the impetus to do something to help. It is important to recognise that activism can take a variety of different forms. O'Brien et al. (2018) present a helpful typology when exploring young people's environmental activism, using the labels dutiful, disruptive, and dangerous dissent:

Dutiful dissent: Young people's concerns are voiced through existing political and social institutions. This might involve joining or initiating groups that work within existing structures and hierarchies. Through participation, young people may engage in political processes and liaise with political elites.

Disruptive dissent: Young people who feel that nothing is being done about their concerns (Della Porta and Diani, 2006) may question, or seek to change existing political, economic and social structures, rules and norms. Action might include petitions strikes and boycotts as well as protest marches and disruption of climate summits.

Dangerous dissent: According to O'Brien et al. (2018, p. 42), this type of action involves "political activism that defies business as usual by initiating, developing, and actualising alternatives that inspire and sustain long-term transformations."

When it comes to something as important as the future of the planet, Demaria et al. (2013) suggest dissent and disruption may be necessary to galvanise the support needed. Indeed Mackey, (2012) points out young people having a say in environmental issues is a fundamental children's rights issue, enshrined in Article 12 of the United Nations Rights of the Child (United Nations General Assembly, 1989).

Compassionate care for the planet

Not all environmental activists will take the form of dissent. We recommend that the dissent typology of young people's activism be complemented by further elements: compassionate care for the planet; congruent living; courageous conversations and community collaboration.

The idea of compassionate care used in an environmental context is not new. Championed by Karen Armstrong (TED, 2008), compassionate care for the earth recognises the interconnectedness of people, animals, and the natural world. It advocates living in harmony with the environment and adopting practices which support sustainability. For children and young people this could mean supporting bio-diversity, eating more plant-based foods in place of meats and finding ways to protect species and habitats.

Congruent living

According to Carl Rogers, congruence is when a person's ideal self is consistent with their actual lived experience. A person is congruent when their actions align with their beliefs and values (Rogers, 1959). This is particularly pertinent in relation to the environment as so often beliefs and actions contradict each other, resulting in cognitive dissonance (the mental discomfort that arises from holding two contrary opinions at the same time). An example of this dissonance could be when an individual truly believes that we must reduce our consumption of consumables to help the planet, yet finds themselves needing to keep up with fashion trends and therefore purchasing several "fast fashion" items every season. Helping children to articulate their beliefs and values about the planet, and then live congruently is a great way to reduce eco-anxiety. For example, it may mean highlighting the link between scarce natural resources and the products that we buy and encouraging children to live more sustainably, reducing consumption, making use of second-hand products, buying well, and buying once.

Courageous conversations

In 2021, a collective of psychologists and academics ran a large-scale study into young people's mental health and the climate crisis. The findings were startling, 75% of the 10,000 participants felt that the future was frightening. However, 48% said they were not listened to when they voiced their concerns (Marks et al., 2021). Having open, honest, and age-appropriate conversations about existential issues such as the climate emergency is key to young people's wellbeing. Beginning this dialogue can be difficult both for children, and the adults that support them. It requires trustful relationships, high quality information and tools to help children manage their emotions and develop self-efficacy.

Community collaboration

There is a prevalent discourse that young people are more engaged in environmental issues than other generations. However, as Ojala (2018) points out, this argument is overly simplistic and often divisive. She identifies that despite their evident planetary concern, many young people's lifestyles are no more sustainable than their parents and grandparents. She argues against "piecemeal behaviour change" (Ojala, 2018, p. 15) in favour of real dialogue within and between communities, where we can ask the difficult questions explore what world we want to live in and what we need to do to achieve this. There is a recognition here that real change will need everyone to work together. For CaYP engaging in environmental action, this might involve allying with concerned others of different generations to maximise impact.

We contend that engaging in environmental action, however small, can help children to experience a sense of agency, and to feel that they are making a difference. Their positive actions can help reduce feelings of helplessness and anxiety, by channelling those emotions into positive and purposeful efforts to create change. Participating in eco-activism can lead to tangible outcomes and positive changes in the environment, and seeing the impact of their efforts can be gratifying and empowering for individuals, providing a sense of hope and optimism for the future. Eco-activism also helps to raise awareness about environmental issues and encourages others to take action. This can help spread information, mobilise support and create a greater understanding and urgency around environmental challenges within communities? With any eco-activism there are likely to be obstacles and setbacks. Overcoming these challenges encourages individuals to develop important skills such as problem solving, collaboration, and perseverance. This growing resilience can help them better cope with eco-anxiety and other challenges that they may encounter in life.

The Case Study below (drawn from email discussions with the young people involved and with permissions in place to use the material) shows how eco-activism has been achieved by a group of young people in a town in Wales. Alice, Daniel, and Ewan live in Pontypridd in South Wales. These young people have engaged in a mixture of dutiful and disruptive activism (O'Brien et al., 2018), as well as compassionate actions, congruent living, and community collaboration. Below they talk about their concern for the environment and the activism that has sprung from this.

Case study – Young people's activism

We are part of a group of young people from Pontypridd in South Wales who are very passionate about the environment. We are fortunate to live in a beautiful part of the world and we are keen to preserve this beauty for future generations.

After being in primary school together for several years, we came together to support biodiversity in our local community. As a group, we went to the school strike for climate protests and became part of the worldwide struggle for the environment. These strikes built our passion for the world, and we took that inspiration home.

We created an organisation called Young Friends of The Earth, Pontypridd so that we could make a difference. This organisation helped us find others who cared for the environment and our group grew in numbers. As part of this group, we were able to help our community. We decided to organise a vegan cake sale to raise money. We raised about £150 in this cake sale and decided to put this money towards plants to put in public places so that we can support pollinating insects while encouraging green spaces in Pontypridd.

We teamed up with others to organise a protest through the streets of Pontypridd in September 2019 as part of a day of global action, and over 300 people of all ages and backgrounds came, including a whole class of local school children. However, during the lockdown, it was impossible to organise events, so our plans were stopped. As young adults we are part of the Youth Forum where we get taken seriously and our voices are heard. As members of this forum, we worked with NATURPonty and local artists to spread awareness for the cause. We held an art exhibition in the Calon Taf centre in Ynysangharad park to showcase some of this work with the local community.

Our activism has helped us to develop new skills, build relationships with others, and feel like we are making a positive difference. Now we just need more people to take up the cause and help us build a more secure future for everyone.

How settings can support children's actions

Up until this point, we have considered activism from the point of view of individuals or small groups. However, educators can play an important role in embedding eco activism into pedagogies, curricula, and systems. Mackey (2012) highlights the importance of giving children good quality information on social and environmental issues and facilitating their right to be part of the environmental conversation and to have their ideas and contributions heard and valued. The case study below (drawn from an interview with the Headteacher) shows how one school adopted a systematic and multi-faceted approach to helping children learn about the environment and explore the difference their own actions could make. Victoria Voller is Headteacher of Shere Infant School in Hampshire and has experienced a growing concern about how best to support children to understand and respond to environmental issues. Below Victoria reflects on her own journey, and that of the school, as she worked with others to embed an ethos of environmental understanding and care for the planet within the school community.

Case study: a whole school approach

As a head teacher, my interest in climate change comes from what the children and parents in the setting were saying to me. Parents were talking about that feeling of powerlessness in relation to the global emergency and the feeling that whatever they did was small and insignificant. All the time, I was hearing "My impact is nothing. And whatever I do cannot change anything." The children were affected too, in a very childlike way. They were hearing about climate issues on television, not fully understanding, and just being really scared about the unknown enormity of it. Some of the children were seeing iconic environmental images like the turtle in a carrier bag and becoming upset. I started to wonder how I could help, what we could do from an education perspective and [I was] inspired by Greta Thunberg, I realised that activism was the way to go. So, I was responding to the children and parents rather than initiating something myself. That is important, I was listening to the voice of the children and the voice of the community and just considering how we could respond as a setting. My setting is built on activism. It was founded in 1842 by two women against the wishes of the church. They wanted to provide education for agricultural workers. Their motto which we still use today was "not like an ordinary school." We base our curriculum around natural pedagogies, and our vision is based around the parable of the mustard seed. At Shere School we have identified six key pedagogies known throughout the school community aa the natural pedagogies: these are play, games, dialogue, nature, stories, and the arts. These pedagogies have served humanity for thousands of years and are designed to mirror children's natural ways of leaning and to respect individual child development. Teachers use these pedagogies to facilitate learning across the curriculum.

So, if you think about our school historically as a place of innovation, then you think about our pedagogies, combined with that ethos of the mustard seed-that small things can be powerful, that small children are valuable, and that small actions matter, we were well placed to take forward an eco-agenda. The school governors and teaching staff were fully on board helped drive this approach forward.

We started with equipping the children with the knowledge to understand what was happening in a child appropriate way, to try and reduce the fear and to make it manageable. After this we started promoting what we call courageous advocacy. We introduced the concept of Fridays for Future and regularly came off timetable and focused on eco issues. The children loved it. We invited in visitors such as beekeepers to talk to the children about ecosystems. We looked at how we could use art creatively, made Christmas decorations out of recycled things, and learned screen printing. We did litter picks in the local community. They were small actions, but the impact of these over time would be huge. We learned about Greta Thunberg and other young activists and even did our own school strike. At this point, parents started telling me how the children were driving their whole families to more sustainable ways of living, they were becoming massive advocates of the earth, really!

We were lobbying BT about a local piece of land and had not heard anything so all the children wrote individual letters, saying what a difference the land would make to the children's learning. Within a week we were given a tenancy at no cost, to create a nature garden. The children found that their voice mattered, and they could be agents of change. It was a powerful lesson.

During Covid, families took it in turns to clear a forest that was gifted to us, and we are now in the process of creating a forest school with a composting toilet and a fire circle. The children spend time there exploring their science curriculum and enjoying nature.

The children were also proud when we received our Eco school green flag. It was a great way to bring together all that we have been doing as a school and to showcase it in a positive way. I have learned from the children's courageous advocacy too, and I am now more confident in championing the school and community and pressing government for the changes we need in education.

This inspiring case study shows what can be achieved with a whole school approach and with a Headteacher, governing body and parent community that is fully on board with the agenda. In schools where this is not the case, educators may want to use the mustard seed approach and start slowly introducing eco approaches and then evaluating their benefits to make a case for further investment and time. Many successful movements of change have begun in just this way.

Conclusion

Climate change is a highly emotive issue for children, young people, and adults alike. The worries that climate emergency elicit are not hypothetical; they are real and are an understandable response to the existential threats we are facing. However, living in fear is not helpful for children and young people's wellbeing. Instead, we want to help them move beyond worry into a context where they are able to contribute positively to the sustainability agenda and live with hope for the future. When a child is experiencing anxiety, there is much that we can do immediately to support them, and here we have explored a range of approaches to help manage anxiety and promote wellbeing. However, in the longer term we want CaYP to feel empowered to play an active role in conversations and actions designed to help realise those hopes of a positive future. This chapter has suggested approaches and shared case studies that show how we, as educators, can help CaYP to help the planet.

Acknowledgements

The authors would like to thank Shere Infant School and Young Friends of The Earth, Pontypridd for their input into this chapter.

References

Ahmed, A., Devi, R.G. and Priya, A.J. (2021) 'Effect of Box Breathing Technique on Lung Function Test', *Journal of Pharmaceutical Research International*, 33(58A), pp. 25–31. Available at: https://doi.org/10.9734/jpri/2021/v33i58a34085.

American Psychiatric Association (2022) Anxiety disorders. *Diagnostic and Statistical Manual of Mental Disorders. 5th ed. Text Revision*. American Psychiatric Association, 2022, 19(2) pp. 215–231.

Armstrong, K. (2009) *My wish: The Charter for Compassion, Ted.com*. TED Talks. Available at: https://www.ted.com/talks/karen_armstrong_my_wish_the_charter_for_compassion. (Accessed 22 February 2025).

Baudon, P. and Jachens, L. (2021) A scoping review of interventions for the treatment of eco-anxiety. *International Journal of Environmental Research and Public Health*, *18*(18), p. 9636.

Chauzy, I. (2020) Seven young eco-activists we should be talking about. *Euronews*. Available at: (accessed https://www.euronews.com/green/2020/08/18/move-over-greta-7-young-activists-you-didn-t-know-about. (Accessed 22 February 2025).

Chawla, L., Keena, K., Pevec, I. and Stanley, E. (2014) Green schoolyards as havens from stress and resources for resilience in childhood and adolescence, *Health & Place*, 28, pp. 1–13. Available at: https://doi.org/10.1016/j.healthplace.2014.03.001. (Accessed 22 February 2025).

Cheng, J.C.H. and Monroe, M.C. (2012) Connection to nature: Children's affective attitude toward nature. *Environment and Behavior*, *44*(1), pp. 31–49.

Clayton, S. and Karazsia, B.T. (2020) Development and validation of a measure of climate change anxiety. *Journal of Environmental Psychology*, *69*, p. 101434.

Coffey, Y., Bhullar, N., Durkin, J., Islam, M.S. and Usher, K. (2021) Understanding eco-anxiety: A systematic scoping review of current literature and identified knowledge gaps. *The Journal of Climate Change and Health*, *3*, p. 100047.

Creswell, C., Waite, P. and Hudson, J. (2020) Practitioner review: Anxiety disorders in children and young people–assessment and treatment. *Journal of Child Psychology and Psychiatry*, *61*(6), pp. 628–643.

Creswell, J.D. (2017) Mindfulness interventions. *Annual Review of Psychology*, *68*, pp. 491–516.

Della Porta, D. and Diani, M. (2006) *Social movements: An introduction*. 2nd edition. Oxford: Blackwell.

Demaria, F. Schneider, F., Sekulova, F. and Martinez-Alier, J. (2013) What is degrowth? From an activist slogan to a social movement. *Environmental Values*, *22*(2), pp. 191–215. Available at:https://doi.org/10.3197/096327113x13581561725194 (Accessed 22 February 2025).

Dippel, A., Brosschot, J.F. and Verkuil, B. (2024) Effects of worry postponement on daily worry: A meta-analysis. *International Journal of Cognitive Therapy*, *17*(1), pp. 160–178.

Dolan, N., Simmonds-Buckley, M., Kellett, S., Siddell, E. and Delgadillo, J. (2021) Effectiveness of stress control large group psychoeducation for anxiety and depression: Systematic review and meta-analysis. *British Journal of Clinical Psychology*, *60*(3), pp. 375–399.

Gladwell, V.F., Brown, D.K., Wood, C., Sandercock, G.R. and Barton, J.L. (2013) The great outdoors: How a green exercise environment can benefit all. *Extreme Physiology & Medicine*, 2, pp. 1–7.

Greimel, K. V. amd Kröner-Herwig, B. (2011) Cognitive behavioral treatment (CBT) In: Greimel, and Kröner-Herwig (eds). *Textbook of tinnitus*. New York: Springer. pp. 557–561.

Hart, A., Gagnon, E., Eryigit-Madzwamuse, S., Cameron, J., Aranda, K., Rathbone, A., & Heaver, B. (2016). Uniting Resilience Research and Practice With an

Inequalities Approach. SAGE Open, 6(4). Available at: https://doi.org/10.1177/2158244016682477

Hickman, C., Marks, P., Pihkala, E., Clayton, S., Lewandowski, E.R., Mayall, E.E., Wray, B., Mellor, C., and Van Susteren, L. (2021) Climate anxiety in children and young people and their beliefs about government responses to climate change: A global survey. *The Lancet Planetary Health*, *5*(12), e863–e873. Available at: https://www.thelancet.com/journals/lanplh/article/PIIS2542-5196(21)00278-3/fulltext (Accessed 22 February 2025).

Hodgkins, A. and Prowle, A. (2023) *Strength-based practice with children and families*. St. Albans Critical Publishing.

Hofmann, S. G. & Gómez, A. F. (2017) Mindfulness-based interventions for anxiety and depression. *Psychiatric Clinics*, *40*(4), pp. 739–749.

James, A.C., Reardon, T., Soler, A., James, G. and Creswell, C. (2020) Cognitive behavioural therapy for anxiety disorders in children and adolescents. *Cochrane Database of Systematic Reviews*, *11*(11). Available at: https://doi.org/10.1002/14651858.cd013162.pub2.

Kandola, A. and Stubbs, B. (2020) Exercise and anxiety. *Physical Exercise for Human Health*.12(2) pp. 345-352. doi: 10.1007/978-981-15-1792-1_23. PMID: 32342469.

Kondo, M. and Shimada, S. (2015) Serotonin and exercise-induced brain plasticity. *Neurotransmitter*, *2*(e793), pp. 10–14800.

Kurth, C. and Pihkala, P. (2022) Eco-anxiety: What it is and why it matters. *Frontiers in Psychology*, *13*, p. 981814.

Léger-Goodes, T., Malboeuf-Hurtubise, C., Mastine, T. and Paradis, P. (2022) Eco-anxiety in children: A scoping review of the mental health impacts of the awareness of climate change. *Frontiers in Psychology*, *13*(13) pp. 1-21. Available at: https://www.frontiersin.org/journals/psychology/articles/10.3389/fpsyg.2022.872544/full (Accessed 22 February 2025)

Mackey, G. (2012) To know, to decide, to act: The young child's right to participate in action for the environment. *Environmental Education Research*, *18*(4), pp. 473–484.

Marks, E., Hickman, C., Pihkala, P., Clayton, S., Lewandowski, E.R., Mayall, E.E., Wray, B., Mellor, C. and van Susteren, L. (2021). Young people's voices on climate anxiety, government betrayal and moral injury: A global phenomenon. *SSRN Electronic Journal* [Preprint]. Available at: https://doi.org/10.2139/ssrn.3918955. (Accessed 22 February 2025).

NHS (2022) *5 steps to mental wellbeing*, *NHS*. Available at: https://www.nhs.uk/mental-health/self-help/guides-tools-and-activities/five-steps-to-mental-wellbeing/. (Accessed 22 February 2025).

O'brien, K., Selboe, E. and Hayward, B.M. (2018) Exploring youth activism on climate change. *Ecology and Society*, *23*(3). Available at: https://www.jstor.org/stable/pdf/26799169.pdf?refreqid=fastly-default%3Ab4a74a95c38e88a2a651af6768a0fddf&ab_segments=&initiator=&acceptTC=1 (Accessed 22 February 2025).

Ojala, M. (2018) Eco-anxiety. *RSA Journal*, *164*(4), pp. 10–15. Available at: https://medium.com/rsa-journal/eco-anxiety-323056def77f.

Öst, L.-G. and Clark, D. (2013) Cognitive behaviour therapy. Principles, procedures and evidence base. In Öst L.-G. & Skaret E. (eds.) Cognitive behavioral therapy for dental phobia and anxiety. Chichester, West Sussex: Wiley-Blackwell, pp. 91–107.

Oxford Reference (2024). *Overview environmental activism.* Available at: https://www.oxfordreference.com/display/10.1093/oi/authority.20110803095753505 (accessed 13 March 2024).

Prakash, R.S., Whitmoyer, P., Aldao, A. and Schirda, B. (2017) Mindfulness and emotion regulation in older and young adults. *Aging & Mental Health*, *21*(1), pp. 77–87.

Rogers, C.R. (1959) A theory of therapy, personality, and interpersonal relationships: As developed in the client-centered framework. In S. Koch (Ed.), *Psychology: A study of a science. formulations of the person and the social context* (Vol. 3, pp. 184–256). New York: McGraw Hill.

Ruiz-Íñiguez, R., Santed German, M.A., Burgos-Julián, F.A., Díaz-Silveira, C. and Carralero Montero, A. (2020) Effectiveness of mindfulness-based interventions on anxiety for children and adolescents: A systematic review and meta-analysis. *Early Intervention in Psychiatry*, *14*(3), pp. 263–274.

Saunders, D. and Kober, H. (2020) Mindfulness-based intervention development for children and adolescents. *Mindfulness*, *11*(8), pp. 1868–1883.

Thompson Coon, J., Boddy, K., Stein, K., Whear, R., Barton, J. and Depledge, M.H. (2011) Does participating in physical activity in outdoor natural environments have a greater effect on physical and mental wellbeing than physical activity indoors? A systematic review. *Environmental Science & Technology*, *45*(5), pp. 1761–1772.

UNICEF (1989) *UN Convention on the Rights of the Child (UNCRC), UNICEF.* Available at: https://www.unicef.org.uk/what-we-do/un-convention-child-rights/. (Accessed 22 February 2025).

Verplanken, B., Marks, E. and Dobromir, A.I. (2020) On the nature of eco-anxiety: How constructive or unconstructive is habitual worry about global warming? *Journal of Environmental Psychology*, *72*, p. 101528.

Wegner, M., Helmich, I., Machado, S., E Nardi, A., Arias-Carrion, O. and Budde, H. (2014) Effects of exercise on anxiety and depression disorders: review of meta-analyses and neurobiological mechanisms. *CNS & Neurological Disorders-Drug Targets (Formerly Current Drug Targets-CNS & Neurological Disorders)*, *13*(6), pp. 1002–1014.

Wolch, J.R., Byrne, J. and Newell, J.P. (2014) Urban green space, public health, and environmental justice: The challenge of making cities 'just green enough.' *Landscape and Urban Planning*, *125*, pp. 234–244.

3

CLIMATE MIGRATION AND DISPLACEMENT

A children's rights perspective

Alison Prowle and Janet Harvell

Introduction

We live in challenging times. Across the world, increasing numbers of children are being forced to leave their homes in search of safety and a more secure future. Escalating conflict in several regions of the world, along with fear of persecution, abnormal weather events and climate change have pushed the numbers of forcibly displaced people worldwide to a record high of 114 million, and more than a third of these are children (United Nations 2023). While the relationship between climate change and climate-related migration is complex and multi- faceted, it is probable that the accelerating climate crisis will cause unprecedented levels of dispacement in the years and decades ahead. It is also important to recognise that this displacement is taking place at a time when there are reduced resources and limited political appetite in less affected countries to host large numbers of refugees.

While forced migration is an extremely challenging experience for all those affected, the contexts and experiences of displaced children vary vastly, depending on a range of individual, family and societal factors. Our experience has shown that some children will find new homes in supportive host communities. They will have their needs understood and met, and will be encouraged to participate fully in their new environment, going on to enjoy healthy and successful futures. Others may experience challenges such as language barriers, hostility and discrimination in host countries, thus exacerbating existing trauma and stress. However, countless other children born into turbulent situations will remain displaced for the remainder of their lives, facing dangerous and precarious futures, without secure access to food, shelter, healthcare and education. It is upon this latter group of children that

DOI: 10.4324/9781003506850-6

this chapter focuses. It is our premise that we need to radically re-imagine education for displaced children, asking some fundamental questions about the purpose of education in this context, how it aligns with the aspirations of displaced children and communities, and how it should be delivered. To this end, while highlighting the harsh realities for children growing up in refugee camps, we are also keen to explore hopeful approaches that could make a positive difference to their lives.

We were privileged to spend time at the Grande-Synthe Camp (Camp de la Linière) in Dunkirk, which opened in 2016 and became one of the first refugee camps in Europe to meet UN standards. Discussions throughout this chapter are informed by these visits and relevant literature.

Climate migration

Climate change migration refers to the forced movement of people from their homes, due to the adverse impacts of climate change including rising ocean levels, extreme weather events, extended droughts, and other environmental changes. As global temperatures continue to rise, the number, frequency of and intensity of such climate-related incidents are increasing. These climate events often exacerbate existing vulnerabilities and inequalities (Climate and Migration Coalition n.d.). Current examples of areas facing climate migration include the following:

- Pacific Islands: Rising sea levels are resulting in significant land loss across the Pacific Islands. Most of the islands' infrastructure sits within 500 metres of the coast and islanders are losing their homes and livelihoods as a result. Without urgent action, the lowest lying islands such as Tuvalu will be completely submerged and uninhabitable within the next few decades. Right now, islanders are challenged by shrinking coastlines, salinisation and shortages of fresh water, ever more intense and unpredictable weather events, ocean acidification and coral reef bleaching (Ober and Waters 2023).
- Bangladesh: Coastal regions of Bangladesh are experiencing increased flooding and salinity of fresh water supplies due to rising sea levels. This has forced indigenous populations to migrate to urban centres to escape the impact on agriculture and livelihoods (Rahman et al. 2019).
- Central America: Drought, crop failures and extreme weather events are leading to economic instability as people migrate towards North America in search of a safer and more secure livelihoods (Booth et al. 2020).
- Sub-Saharan Africa: Desertification and crop failures are resulting in food insecurity and displacement of local populations. Lake Chad has shrunk significantly, impacting on the livelihoods of millions of people who rely on its water and agriculture (Ofori et al. 2021).

- Northern Canada: Indigenous communities are dealing with the effects of melting permafrost, increased frequency of wildfires and rapidly changing ecosystems. This is prompting consideration of relocation or adaptation (Prichard et al. 2021).
- Syria: Prolonged drought and climate variability have contributed to agricultural failures in Syria, leading to food insecurity and displacement of rural populations to urban areas. This, in turn has exacerbated social and political tensions and resulted in people fleeing the country (Daoudi 2020).

As habitable areas become smaller and there is greater pressure on diminishing resources, for millions of people, the only option available will be life in refugee camps. Below we discuss Cox Bazaar, Bangladesh, a refugee camp affected by adverse weather events (Zaman et al. 2020). This example was chosen as it is the largest refugee camp in the world and may, unfortunately, provide a disturbing glimpse of what the future may look like for countless children.

Cox Bazaar in Bangladesh comprises numerous refugee camps and is home to more than one million people. Of these, between four and five thousand inhabitants are children (exact figures are difficult to ascertain owing to a fluctuating population and limited infrastructure). Miks (2024) describes the situation of Rohingya (a stateless Indo-Aryan ethnic group from Myanmar) children living in Cox Bazaar refugee camp:

> Many of the almost half a million children in the camp live in flimsy bamboo and tarpaulin shelters perched on hilly terrain. Particularly during the torrential rains of the monsoon season, the sandy soil in the camps can turn to mud, bringing flooding and landslides even as powerful winds wreak havoc on already vulnerable families' homes.

The children have lived through outbreaks of fire, disease, and a cyclone, all of which have worsened their living conditions and made life even more precarious. Illness is rife in the camps with children's already compromised immunity, leading to frequent coughs, colds and sickness bugs. Over 40% of the children are chronically malnourished. Child labour affects 10% of children, as their situation compels them to work to assist families who are living in absolute poverty. Humanitarian organisations have managed to instigate positive changes, including increasing food rations, and enrolling more children within an expanded education offer. However, any improvement in living conditions is likely to be short lived as Bangladesh is one of the most climate-vulnerable nations in the world, and millions of additional people may, in time, be forced to flee their homes as sea levels continue to rise. This will put additional strain on the already scarce resources of the camps with ongoing adverse impacts on the lives of all who live there.

Understanding the educational context

Article 28 of the United Nations Convention on the Rights of the Child (UNCRC 1989) states that all children and young people have a right to education. This is the case, regardless of race, gender, disability, detention or refugee status. However, there are a number of challenges that children may face in relation to accessing this right. These include:

- Disruption to education: In their home country, war, conflict or disaster may have made it difficult for public services to operate, resulting an education that is fragmented and sporadic. During transit, more education time may be lost and when arriving in host countries, a child's education may be disrupted by frequent moves and lengthy delays in accessing provision (Prowle 2022). UNHCR suggests that on average, refugee children miss out on three to four years of schooling owing to such disruptions.
- Trauma: Trauma is an emotional response to a terrible event such as witnessing violence or living through a natural disaster (Babbel 2010). For children forced to leave their home, these events might be profound, prolonged and multiple (Rankin and Regan 2004; Prowle and Musgrave 2018). Initial trauma responses may include shock and denial. However, longer term effects of trauma may include ongoing anxiety, Post-Traumatic Stress Disorder (PTSD), hypervigilence or excessive risk behaviours, relationship difficulties, and somatic symptoms. Responsive psychosocial support will be necessary to support the young person. Unfortunately, education providers are often ill-equipped to provide what is needed (McDiarmid et al. 2022).
- Access to education: This may be due to the absence of an infrastructure for supporting refugee children in host communities or refugee camps. This is often hampered by a lack of resources including learning materials and staff.
- Language barriers: Differences between the refugee children and the staff/volunteers that support them may hinder children's ability to fully engage within learning.

Addressing these issues requires an holistic approach that considers the needs and challenges faced by refugee children and provides support to overcome them. However, it is also about building trustful relationships with the young person (Hynes 2017), recognising their uniqueness and potential (Kohli 2018), and providing a trauma-aware environment that promotes a sense of emotional safety (Prowle 2022). While there are excellent examples of provision both in host countries and in refugee camps, it is not consistent and there are limited opportunities for sharing good practice. The statistics are shocking; UNICEF has estimated that globally 51% of refugee children are not in school. Enrolment varies dramatically between phases, with approximately 42% accessing a secondary education and 6% and 3% respectively accessing tertiary or higher education.

Education can be transformative for children's lives and their life chances (Margaroni and Magos 2018). It can reduce inequalities in society and provide displaced children with the skills, knowledge and empowerment needed to pursue future economic opportunities, break cycles of poverty and make a positive contribution to rebuilding communities (Naseh et al. 2024). At best, education can break down barriers and foster mutual understanding between refugees and local populations, thus promoting inclusion (Dryden-Peterson et al. 2018). It can promote better health and wellbeing outcomes for displaced children by providing them with access to information and resources to support their own wellbeing. Educational programmes can offer safe and structured environments for displaced children, protecting them from exploitation violence and other risks (Sengupta et al. 2019). It can provide a supportive environment helping displaced children to cope with trauma, build resilience and develop positive relationships with peers. Conversely, disruptive, or absent education can serve to reinforce inequalities and compound divisions. UNESCO (2016) conducted a longitudinal research project and found that low levels of education heighten the risk of violence and conflict, creating a cycle of lost opportunities, ongoing strife, and displacement.

Children's lives in refugee camps

The United Nations High Commissioner for refugees defines refugee camps as:

> temporary facilities built to provide immediate protection and assistance to people who have been forced to flee their homes due to war, persecution or violence. While camps are not established to provide permanent solutions, they offer a safe haven for refugees and meet their most basic needs such as food, water, shelter, medical treatment and other basic services during emergencies.
>
> *(UNHCR 2024a*

The continuing need for refugee camps remains critical as the growth of displaced people is set to increase as new conflicts, climate change and natural disasters occur at an unprecedented rate. Indeed, in 2023 the UN identified that the number of refugees had reached 110 million people, the largest increase in a single year, with approximately 6.6 million living in refugee camps. It is estimated that there are currently 500 refugee camps globally, with the average camp home to 11,500 refugees. However, the world's five largest camps are home to 201,000 (Kakuma), 240,000 (Dadaab), 83,000 (Za'atari), 9300,000 (Kutupalong) and 16,890 (Um Rakuba) respectively.

As UNHCR (2024) explained, by their nature, refugee camps were intended as temporary places of refuge, but ongoing wars and natural disasters have meant that many of the larger camps have become permanent fixtures, with refugees spending years, decades and even generations living there. The largest of these camps have even formed their own urban networks with schools, functioning drainage systems, markets and even speed bumps in some of the more established sites (Filipenko 2023). Common challenges that the inhabitants of these camps face include: coming to terms with separation from family and loved ones and the resulting loss of community; a lack of food, water, and shelter; limited or no access to schooling and fears about the future.

On these camps, overcrowding can result in basic infrastructures being overwhelmed as key supplies and services such as food, water, and health care become inadequate to meet the needs of the refugees. Indeed, during our visits to the Dunkirk camp in France, we would observe queues for food at mealtimes, and on occasions the food had run out before everyone had eaten. As women and children make up approximately 60% of the refugee population, it is inevitable that they are disproportionately affected and most vulnerable to effects of the displacement and trauma experienced. As we observed during our visits to the French camps, the impact on children's emotional and social development is of particular concern, especially when one combines this with the risks of abuse, trafficking, violence, and exploitation which can be experienced during the refugee journey, as well as once settled in the camps (UNHCR 2022; UNODC 2022).

Similarly, the effect of restricted and unhealthy dietary options has, in some instances, resulted in undernutrition, such as wasting, stunting, and deficiencies in vitamins and minerals (Filipenko 2023). This is a particular worry when considering the importance of healthy diets in the early years of children's lives and the impact that such deprivations could have on their development.

Accessing schooling can also be challenging, with only 68% of children attending primary school and 34% secondary school, compared to 91% and 84% worldwide respectively (UNHCR 2021). The quality of this education is also problematic. A lack of appropriate buildings and resources add to the problems and local issues are raised by those organisations/local authorities managing the camps. This was something that we observed during our own visits to the Dunkirk camp when children were bussed to local schools for part of the day. These classes had no interpreters and children were expected to sit in the classes with little language support. As a result, it was noticeable that school-aged children disappeared from public areas before school bus collections, 'voting with their feet.'

We also observed boredom as a significant challenge for those living in refugee camps, with little to break up the day. This is possibly one of the biggest challenges given its ripple effect on mental health and is a contributing factor to the anti-social behaviours such as gang culture and violence that can occur. For many, their day is spent on mobile telephones keeping in touch with family or people smugglers. Different charity organisations will also provide opportunities to engage in art, music, or language sessions, but for most their time in camps is just a 'waiting game.'

Children are at particular risk, being more likely than adults to die of natural hazards in addition to being more susceptible to diseases, malnutrition, or injuries (Vaghri et al. 2019). Worryingly, many countries that are particularly vulnerable to climate change also have populations with significantly higher numbers of children, and increasing temperatures impact the health of children disproportionately. Children are more vulnerable due to their lower tolerance for heatwaves, air pollution, and food deprivation/undernutrition resulting in high child mortality. Eighty-eight percent of the existing liability of diseases because of climate change impacts children under the age of five (Climate Analytics 2019).

But children are not only at physical risk, the psychological impacts of extreme events can also result in emotional distress for children including fear of separation from their families. A study conducted in southeast USA following Hurricane Andrew and another study in Poland conducted after the 1997 floods showed that children are two to three times more likely to suffer from post-traumatic stress disorder than adults (La et al. 1996).

During our visits to the refugee camps it was clear that most refugees spent their time planning to travel to the UK, largely via people smugglers arranging for refugees to get onto lorries or dinghies to make the dangerous crossing across the English Channel. As this needed to take place in the dark when there was less chance of discovery, most attempts took place during the night or early morning. When speaking with staff at the camp medical centre, they spoke of common symptoms as a result of these attempts, including burns caused by refugees being forced into cramped conditions against heat sources for several hours during their escape attempts, broken limbs when jumping from the lorries and cumulative mental health issues as a result of their traumatic experiences.

There are positive steps being taken to address these issues, which we explore later under 'Envisioning a hopeful future.' Similarly, there are some exciting interventions to develop healthy habits and address issues of poor diet for refugees. For example, in some of the more established camps, advice, guidance, resources, and training have promoted the development of solar power to access water from boreholes, resulting in the growing of crops for personal use and trading (Filipenko 2023).

Envisioning a hopeful future

The foregoing sections have painted a bleak picture of the realities of children's lives in relation to global migration. Overall, refugee camps are not conducive to children's wellbeing, although they are a better option than homelessness or remaining in immediate danger. While it is acknowledged that education can play an important role in empowering displaced children by providing knowledge skills and opportunities, the reality for many is an education that is fragmented, inadequate and lacks congruence with their current circumstances. In light of the climate emergency, it is important to consider not only how we provide education to displaced children, but also to go back to first principles and ask questions concerning the purpose of education and how it can best meet the needs of children who have already lost much.

Utopia as method

Levitas (2013) suggests that we need to gaze beyond the current realities that already exist, to challenge existing norms and paradigms, and imagine alternative scenarios and possibilities. This is referred to as a Utopian method and provides a way for individuals and communities to envisage and strive towards a better future. Utopia, in this context, is not an unreachable idyll, but an instrument of social change and community activism. This approach is extremely valuable, given the dystopian scenarios we are facing. However, in our Utopian idealising, we do not need to start from scratch, as there are some existing models that provide inspiration when considering what an education for displaced children might look like. Next, we explore three such hopeful approaches.

Aga Khan Foundation

Recognising the challenges faced by disposed families and the countries supporting them in refugee camps, the United Nations Assembly agreed the Global Compact on Refugees framework in 2018, intending to provide "more predictable and equitable responsibility-sharing, recognising that a sustainable solution to refugee situations cannot be achieved without international cooperation" (UNHCR.UK 2024). The framework provides guidance for governments and international organisations on how best to help refugees lead fruitful lives, including non-discriminatory access to national education systems for refugee children. This can be particularly challenging when trying to meet the needs of a wide range of cultures, ethnicities, and faiths. The Aga Khan Foundation (AKF 2020) is one of the world's largest, non-profit development agencies, whose aims are to improve the quality of life for poor and marginalised people. Working in many countries including those with significant refugee populations such as Afghanistan, Kenya, Pakistan, Syria, Tajikistan, Tanzania and Uganda, the Foundation has a particular focus on early childhood development and was instrumental in supporting the Kenyan government's development of the Madrasa Early Childhood Program

(MECP) in the 1980s. This program focused on developing an appropriate curriculum for Muslim children that was both religious and secular, while preserving community values and improving educational outcomes of Muslim children (AKF 2008). A resource kit is freely available to access from the website (www.akf.org.uk) and is recommended for use by governments when supporting refugee children.

In addition to the MECP, the AKF continued, in collaboration with the Kenyan Ministry of Education, to develop a modern curriculum, the Competency Based Curriculum (CBC), designed to meet current and future needs of the country. This curriculum was designed to meet the academic, social, emotional, and physical elements of children's learning across twelve learning areas, with an option for two additional subjects (AFK 2021). It aims to equip students with diverse skills needed for the 21st-century workforce, focusing on nurturing the talents, interests, values, and abilities of children. Introduced in Kenya in 2017, the AGF together with the UNHRC and Kenyan Ministry of Education also adopted this within the Dadaab and Kakuma refugee camps, giving refugees access to skills and qualifications promoted by the GCR. Initially introducing early years education and the first three years of lower primary, it is hoped to extend this further as resources become available. There are challenges, key being sufficient, qualified teaching staff in a profession that is already challenged by high turnover rates, limited infrastructure, and the poor learning environments. Nevertheless, the content and structure of the curriculum provides a structure and content that respects the culture and beliefs of the families, while providing a secure educational foundation for the 21st century (AFK 2021, Sections 2.2–2.3).

Amala education

The Arabic word *amala* signifies hope, a fitting name for an organisation conceived to address the gap in high quality education for young people forced to flee their homes. Aimed at young people aged 16–25, Amala has developed a global high school diploma that was designed with refugee young people's needs at its centre. Completing this diploma provides opportunity for refugee young people to complete their formal education, which may have been sporadic, disrupted or non-existent up to that point. It recognises and addresses the barriers that displaced young people face, offering a flexible blended programme The delivery includes of face to face and online elements, as well as trained facilitators from the local area – many of whom are Amala graduates and have a lived experience of the challenging contexts in which the young people are studying.

Both the curriculum and the pedagogies of Amala are centred around building young people's agency (OECD 2018) and developing higher level competencies that will help the young people navigate their way through changing circumstances and environments (Grayling 2017). The curriculum aims to be transformative, both student's own life and experience, but also of society, instilling in students a concept of themselves as agents of change. The OECD's (2018) Transformative Competencies of: Creating New Value (innovation), Reconciling Tensions and Dilemmas (Problem solving and emotional intelligence) and Taking Responsibility (agency and leadership) underpin the modules studied by students, while personal interest projects (PIPs) allow students to pursue learning that is of particular relevance to them as individuals or communities. Pathway Advisors supplement the taught and self-directed aspects of the programme, providing advice, coaching, and mentoring to help the young people achieve their goals (Amala 2023).

Recognising that not all students will be ready or able to commit to the diploma (which requires an average of 15 months to complete), Amala have also developed a series of 10 Changemaker courses, which allow students to return to learning and develop skills for life and study. These courses range from Social Entrepreneurship to Creative and Cultural Expression or Living Peacefully and are geared towards enabling young people to develop knowledge and skills, to support themselves, and to contribute to their communities (Amala Education n.d. in refs).

While this approach is still very new and relatively small scale (in 2023 Amala reported worked with approximately 15,000 young people, through provision in Kenya, Jordan, Uganda, and Greece during the previous year), it does highlight the value of working collaboratively with communities and of offering both short courses aimed to re-engage young people with education as well as a recognised high school qualification. It also underscores the importance of curriculum and pedagogies that recognise young people's starting points, acknowledging, and valuing their lived experience and providing learning that is meaningful in both their current contexts and aspired futures. Challenges to scaling up this approach would be training programme facilitators and mentors, identifying, and procuring suitable resources and infrastructure and developing the offer beyond its current focus on those with at least some English.

The Barefoot College

The Barefoot College, established in an Indian village in 1972 by "Bunker" Roy, takes a radically different view of education, focusing on traditional skills and knowledge, and creating learning to meet real world challenges encountered in rural contexts (Barefoot College 1972. Based upon Ghandi's key principle of self- reliance (Elkington and Hartigan 2008), the college trains individuals in skills and solutions needed by their communities, including irrigation, solar power, and land management. The focus is not on academic achievement; indeed, many college participants are illiterate or semi-literate, but on the effective development of practical skills, and on enabling dignity, empowerment, and self-sufficiency. The wisdom and knowledge of those with ancient skills in medicine, engineering and agriculture is harnessed for community benefit and for sharing with others. Hence community solutions become sustainable. Meanwhile, traditional Indian puppetry has been used to provide education about health, literacy, and human rights.

Focusing especially on women and girls, the Barefoot College works with individual communities to ascertain what the needs are and then to collaboratively develop a programme of learning to meet those needs. The pedagogy is as much about developing sustainability, upending social hierarchies, and empowering those who are marginalised, as it is about transmitting practical knowledge and skills (John 2003). As such, it is an example of community development at its best, winning numerous international awards and accolades for its work and global impact.

To date, the model has not been targeted to working with displaced populations. However, its grassroot approach, focus on empowerment and ability to make use of existing expertise within the camps, suggest that this module model could be easily adapted to provide solutions for displaced communities, especially those living in under-resourced environments such as refugee camps.

Conclusion: towards a sustainable model of education for displaced children

This chapter has noted the significant challenges that lie ahead for the education of the growing number of children displaced by climate change. As educators, this calls us to re-think the purpose and nature of education, if it is to support displaced children and their communities to rebuild their lives, recover from trauma and create new opportunities for the future.

The hopeful approaches identified above provide a glimpse of what might be possible in addressing the right to education of children displaced by war, conflict, or climate events. The projects outlined provide useful principles to underpin our Utopian ideation, these include:

- Recognising the lived experiences of displaced children
- Using culturally aware and trauma-informed approaches
- Teaching what is relevant and helpful
- Making use of resources and individuals within the community
- Ensuring that children's rights underpin the curriculum and pedagogies
- Making use of technological solutions and blended approaches.

Implementing such approaches will not be easy, there are many structural issues that need to be addressed to scale up existing projects, share good practice and ensure consistent quality and sustainability. However, as the successful projects above highlight, where there is a will, there is often a way.

References

Aga Khan Foundation (2021) Refugees to Benefit from New Education Programme. Available at: https://the.akdn/en/resources-media/whats-new/in-the-media/refugees-benefit-new-education-programme (Accessed 9 February 2024).

AKF (2008) Madrasa Early Childhood Program. Available at: https://akflearninghub.org/documents/madrasa-early-childhood-development-programme/ (Accessed 10 February 2024).

AKF (2020) About Us Available at: https://www.akf.org.uk/who-we-are/about-akdn/#:~:text=AKDN%20has%20agreements%20and%20protocols,%2C%20Pakistan%2C%20Portugal%2C%20Russia%2C

AFK (2021) The Implementation of the CBC in Dadaab and Kakuma Refugee Camps. Available at: https://static1.squarespace.com/static/5cfe2c8927234e0001688343/t/61e14d5fb913f0197bc945e2/1642155387275/Samuel+Hall+-+Implementing+the+CBC+in+Dadaab+and+Kakuma+-+Final+report+-+For+Dissemination+%285%29.pdf (Accessed 12 February 2024).

AKF (2024) About Us. Available at: https://www.akfusa.org/about-us/#:~:text=The%20Aga%20Khan%20Foundation%20partners,of%20people%20from%20all%20backgrounds (Accessed 9 February 2024).

Amala Education (2023) Impact Report 2022–2023. https://www.amalaeducation.org/impact (Accessed 21 February 2024).

Amala Education (n.d.) Change Maker Courses. https://www.amalaeducation.org/short-courses (Accessed 21 February 2024).

Babbel, S. (2010) The Trauma That Arises From Natural Disasters. Natural disasters often result in mass cases of PTSD for survivors. Available at: https://www.psychologytoday.com/gb/blog/somatic-psychology/201004/the-trauma-arises-natural-disasters#:~:text=The%20American%20Psychological%20Association%20defines,reactions%20to%20it%20and%20symptoms. (Accessed 26 February 2025).

Barefoot College (1972) Welcome to the real Barefoot College. Available at: https://barefoot.college/ (Accessed 26 February 2025).

Booth, J.A., Wade, C.J. and Walker, T.W. (2020) *Understanding Central America: Global Forces and Political Change*. Routledge.

Climate Analytics (2019) Global Climate Change Impacts on Children. Available at: https://ca1-clm.edcdn.com/assets/report_cc_impacts_on_children_2019.pdf (Accessed 13 February 2024).

Climate and Migration Coalition (n.d.) Climate Change Could Re-Shape Patterns of Migration and Displacement. Available at: https://climatemigration.org.uk/our-work/ (Accessed 9 April 2024).

Dryden-Peterson, S., Adelman, E., Alvarado, S., Anderson, K., Bellino, M., Brooks, R. and Suzuki, E. (2018) *Inclusion of Refugees in National Education Systems*. UNESDOC Digital Library.

Elkington, J. and Hartigan, P. (2008) *The Power of Unreasonable People: How Social Entrepreneurs Create Markets That Change the World*. Harvard Business Press.

Filipenko, D. (2023) Five Largest Refugee Camps in the World and How People Live There. Available at: https://www.developmentaid.org/news-stream/post/158870/five-largest-refugee-camps-in-the-world (Accessed 12 February 2024).

Global Compact on Refugees (2024) The Global Compact on Refugees. Available at: https://globalcompactrefugees.org/about-digital-platform/global-compact-refugees#:~:text=It%20provides%20a%20blueprint%20for,the%20pressures%20on%20host%20countries (Accessed 9 February 2024).

Grayling, A. (2017) "Creating New Value" Section of Future and Education and Skills 2030: Reflections on Transformative Competencies 2030. OECD. Available at: https://one.oecd.org/document/EDU/EDPC(2017)16/ANN5/en/pdf

Hynes, P. (2017) Trust and Mistrust in the Lives of Forcibly Displaced Women and Children. *Families, Relationships, and Societies*, 6(2), pp. 219–237.

John, M. (2003) *Children's Rights and Power: Charging Up for a New Century*. Jessica Kingsley Publishers. ISBN 9781853026584.

Kohli, R.K. (2018) *Protection: Migrant Children and Institutional Protection*. In: Research Handbook on Child Migration. Cheltenham, UK: Edward Elgar Publishing. pp. 381–396. https://doi.org/10.4337/9781786433701.00038

La Greca A, Silverman WK, Vernberg EM, Prinstein MJ. (1996) Symptoms of posttraumatic stress in children after Hurricane Andrew: a prospective study. IN: J Consult Clin Psychol. 1996 Aug;64(4):712-23. doi: 10.1037//0022-006x.64.4.712. PMID: 8803361

Levitas, R. (2013) *Utopia as Method: The Imaginary Reconstitution of Society*. Springer.

McDiarmid, S., Durbeej, N., Sarkadi, A. and Osman, F. (2022) Schools' and Teachers' Roles and Challenges in Supporting the Mental Wellbeing of Refugee Youths: A Qualitative Study with Swedish Teachers. *International Journal of Qualitative Studies on Health and Well-being*, *17*(1), p. 2007568.

Margaroni, M. and Magos, K. (2018) Refugee Experience and Transformative Learning. *Global Education Review*, *5*(4), pp. 194–210.

Miks J. (2024) Picturing a Future Amid the Climate Crisis UNICEF. https://www.unicef.org/stories/picturing-future-amid-climate-crisis (Accessed 19 February 2024).

Naseh, M., Potocky, M., Burke, S.L., Stuart, P.H. and Huffman, F.G. (2024) Factors Associated with Poverty among Refugees in the United States. *Journal of Poverty*, *28*(2), pp. 91–109.

Ober, K. and Waters, K. (2023) Pacific Island Nations Seek Climate Solutions Outside of COP28 United States Institute of Peace. Available at: https://www.usip.org/publications/2023/11/pacific-island-nations-seek-climate-solutions-outside-cop28 (Accessed 28 February 2024).

OECD (2018) The Future of Education and Skills: Education 2030. Position Paper. Available at: https://www.oecd.org/content/dam/oecd/en/publications/reports/2018/06/the-future-of-education-and-skills_5424dd26/54ac7020-en.pdf.

Ofori, S.A., Cobbina, S.J. and Obiri, S., 2021. Climate Change, Land, Water, and Food Security: Perspectives from Sub-Saharan Africa. *Frontiers in Sustainable Food Systems*, *5*, p. 680924.

Prichard, S.J., Hessburg, P.F., Hagmann, R.K., Povak, N.A., Dobrowski, S.Z., Hurteau, M.D., Kane, V.R., Keane, R.E., Kobziar, L.N., Kolden, C.A. and North, M. (2021) Adapting Western North American Forests to Climate Change and Wildfires: Ten Common Questions. *Ecological Applications*, *31*(8), p. e02433.

Prowle, A. (2022) *'Thank You for Asking Me about My Story': An Exploration of the Perspectives of Forced Migrant Parents, Practitioners, and Strategic Actors in South Wales*. PhD Thesis, Cardiff University.

Prowle, A. and Musgrave, J. (2018) *Utilising Strengths in Families and Communities to Support Children's Learning and Wellbeing*. In: Pedagogies for Leading Practice.

Thinking About Pedagogy in Early Childhood Education . Routledge, London. ISBN Paperback: 9781138577428 Hardback: 9781138577398

Rahman, M.M., Ahmad, S., Mahmud, A.S., Hassan-uz-Zaman, M., Nahian, M.A., Ahmed, A., Nahar, Q. and Streatfield, P.K. (2019) Health Consequences of Climate Change in Bangladesh: An Overview of the Evidence, Knowledge Gaps and Challenges. *Wiley Interdisciplinary Reviews: Climate Change*, *10*(5), p. e601.

Rankin, J. and Regan, S. (2004) Meeting Complex Needs in Social Care. *Housing, care, and Support*, 7(3), pp. 4–8.

Sengupta, E., Reshef, S. and Blessinger, P. (2019) Creating a Borderless World of Education for Refugees. In Sengupta, E. & Blessinger, P (Eds)*Language, Teaching, and Pedagogy for Refugee Education* (Vol. 15, pp. 181–191). Emerald Publishing Limited.

UN (1989) United Nations Convention on the Rights of the Child. Available at: https://www.unicef.org.uk/wp-content/uploads/2016/08/unicef-convention-rights-child-uncrc.pdf (Accessed 26 February 2025).

UN (2018) Global Compact on Refugees. Available at: https://globalcompactrefugees.org/sites/default/files/2019-12/Global%20compact%20on%20refugees%20EN.pdf (Accessed 9 April 2024).

UN (2023) With Conflicts, Climate Change Pushing Displacement to Record Levels, Third Committee Highlights Need to Ramp Up Refugee Funding, Address Root Causes. Available at: https://press.un.org/en/2023/gashc4395.doc.htm#:~:text=in%20feeling%20aggrieved.-,In%20the%20ensuing%20interactive%20dialogue%2C%20delegates%20voiced%20concern%20over%20the,people%20was%20reached%20in%202023 (Accessed 10 February 2024).

UNHCR (2021) UNHCR Report 2021: 'Staying the Course' – The Challenges Facing Refugee Education. Available at: https://www.unhcr.org/media/unhcr-education-report-2021-staying-course-challenges-facing-refugee-education (Accessed 12 February 2024).

UNESC) (2016) Missing Out Refugee Education in Crisis. Available at https://uis.unesco.org/sites/default/files/documents/missing-out-refugee-education-in-crisis_unhcr_2016-en.pdf (Accessed 26 February 2025).

UNHCR (2022) Statement on risks of Trafficking and Exploitation Facing Refugees from Ukraine Attributed to UNHCR's Assistant High Commissioner for Protection. Available at: https://www.unhcr.org/uk/news/news-releases/statement-risks-trafficking-and-exploitation-facing-refugees-ukraine-attributed (Accessed 9 April 2024).

UNHCR (2024b) Children. Available at: https://www.unhcr.org/uk/ (Accessed 9 February 2024).

UNHCR (2024a) What Are Refugee Camps. Available at: https://www.unrefugees.org/refugee-facts/camps/#:~:text=Refugee%20camps%20are%20temporary%20facilities,to%20war%2C%20persecution%20or%20violence (Accessed 10 February 2024).

UNHCR.UK (2024c) The Global Compact on Refugees. Available at: https://www.unhcr.org/uk/about-unhcr/who-we-are/global-compact-refugees (Accessed 10 February 2024).

United Nations (2023) Global Refugee Forum Opens Amid 'Crisis of Humanity'. Available at: https://news.un.org/en/story/2023/12/1144767 (Accessed 10 February 2024).

UNODC (2022) Abused and Neglected: UNODC Study Reveals Violence Suffered by Migrants. Available at: https://www.unodc.org/lpo-brazil/en/frontpage/2022/01/abused-and-neglected-unodc-study-reveals-violence-suffered-by-migrants.html (Accessed 9 April 2024).

Vaghri, Z., Tessier, Z. & Whalen, C. (2019) Refugee and Asylum-Seeking Children: Interrupted Child Development and Unfulfilled Child Rights. *Children (Basel)* 6(11), p. 120.

Zaman, S., Sammonds, P., Ahmed, B. and Rahman, T. (2020) Disaster Risk Reduction in Conflict Contexts: Lessons Learned from the Lived Experiences of Rohingya Refugees in Cox's Bazar, Bangladesh. *International Journal of Disaster Risk Reduction*, *50*, p. 101694.

4

HEAD, HEART AND HANDS

Reimagining holistic education

Kathleen Bailey and Nicola Watson

Introduction

In thinking about how we might work with Children and Young People (CaYP) in ways which will serve them in an unknown future, it is important to understand the philosophical framework upon which our pedagogy is based. The term 'pedagogy,' which we understand as the art and science of teaching, stems from the Greek for 'to lead the child.' This invites us to consider where and why we want to lead children. Pedagogies are context-specific, so the meanings and purposes of our education systems are dependent upon the culture, ideology and imperatives that predominate in place and time. As we have seen in Chapter One, presently, materialistic, consumer lifestyles prevail in our western culture. This is unsustainable given the impact on the planet. Therefore, pedagogues need to consider the questions: What does it mean to lead a good life and what do we want to pass on to the CaYP with whom we work?

Increasingly, commentators on education are recognising the need for a paradigm shift away from the individualistic, market-based ideologies which permeate education systems in the industrialised world. Indeed Stein (2019, p. 6) states that "there is no viable future for civilisation that does not include a radical change in the nature of our educational systems." Although such stark pronouncements might provoke discomfort, confusion or denial, they also invite us to confront our personal and professional identities. They offer the opportunity to reconsider our pedagogy and respond creatively, deepening intrinsic motivation and purpose.

In proposing ways we might navigate the uncertainties of the future in our practice, Gallagher and Stobbs (2023, p. 2) suggest we need to "begin the task of proposing ecologies as yet untried and imagining human natures

DOI: 10.4324/9781003506850-7

as yet unseen." This rightly frames the pedagogue as an activist, one who can make a difference in changing the course of the future. However, Moss and Cameron (2020, p. 221) caution that future oriented practice potentially "prioritises a certain idea of what children should become at the expense of what is valuable and meaningful to children in the present." In addressing this tension, we suggest that pedagogies need to be grounded in co-creation and collaboration with CaYP.

In what follows, we consider what a Head, Heart and Hands (HHH) pedagogy might look like, if we were to expand the concept to include the more than human, and so embrace a truly holistic view of the world. In response to Gallagher and Stobbs' (2023) suggestion, we draw on the post-human paradigms of indigenous cosmologies and new-materialist ideas, to imagine a pedagogy that is inextricably entangled with the more-than-human world; one which demands an ethic of care (Barad, 2007). Tronto (1993, p. 103 cited Puig de la Bellacasa, 2011) describes care as:

> ...everything that we do to maintain, continue and repair 'our world' so that we can live in it as well as possible. That world includes our bodies, ourselves, and our environment, all that we seek to interweave in a complex, life sustaining web.

An ethic of care is therefore, a relational, contextual approach which respects the voices of all parties, as we exemplify in the case study later in this chapter.

Post-humanism: reimagining nature

It is a useful starting point to consider what it means to 'connect' with nature. For post-humanist scholars, nature and culture are understood as inextricably entwined; one cannot exist without the other (Latour, 2017). Latour (2017) suggests that nature is a construct created by a humanist understanding of the world, with the human (the archetypal white western male) positioned outside of, and superior to the rest of the world. As such, a nature/culture divide is created, which Latour (2017) terms, the 'Great Divide.' In rejecting this divide, we adopt Puig de la Bellacasa's (2017, p. 1) phrase of 'the more-than-human-world' since it "speaks in one breath of nonhumans and other than humans such as things, objects, other animals, living beings, organisms, physical forces, spiritual entities, and humans." The intention is to decentre the human as the only one who can know, thus opening up the practice of knowledge production to the more than human (Bailey, 2024; Ulmer, 2017). The feminist physicist Barad (2007, p. 185) suggests, "We don't obtain knowledge by standing outside the world; we know because we are of the world." In accepting that we are inextricable from the more-than-human world, we are ethically implicated in what the world becomes (Barad, 2007).

The origins of head heart and hands

The principle of HHH was conceived by Johann Pestalozzi (1746–1827), whom Soëtard (1994, p. 297), describes as "the father of modern educational science." Pestalozzi argued that the role of the pedagogue was to support the development of CaYP in ways that enabled students' independence, by combining the functions of the head, the heart and the hands: the head to develop their own ideas; the heart in forming mutually beneficial relationships in cooperation with others; the hands in realising their learning through physical work. Central to the principle was the idea that the three aspects should be kept in equilibrium so that students developed "into adults who were self-fulfilled, purposeful and autonomous" (Sellars and Imig, 2021). HHH represents an early example of a holistic pedagogy; "the belief that the parts of something are intimately connected and explicable only by reference to the whole" (Trodd, 2016, p. 493).

It is worth noting that Pestalozzi's ideas are underpinned by Rousseau's (1712–1788) conceptualisation of 'nature's child' (Taylor, 2013). Responding to what he saw as the evils of rapidly expanding industrialisation, alongside a reimagining of nature as attractive and pure, Rousseau proposed a template for an ideal society (Taylor, 2013; Urry, 2000). Significantly, Rousseau drew on the emerging nature/culture divide, aligning children with the purity of nature. He suggested that children be raised in and by nature until the age of thirteen; lest they be corrupted by adult society (Latour, 2017; Taylor, 2013). Taylor (2013, p. 10) suggests that this view has become institutionalised in ECEC today via the pedagogies of Froebel, the Macmillan sisters, Montessori, Steiner and Reggio Emilia. The recent interest in ECEC in outdoor pedagogies such as Forest School, also reflect the concept of 're-connecting' children to nature (Bailey, 2020).

Problematically, while 'nature's child' is viewed as instilled with an innate knowledge that must be nurtured by nature, Prout (2005) suggests that in contemporary society, the child's body and its development has become the focus, and the defining characteristic, of childhood, creating a second construction of the child as 'naturally developing.' This construction emerged from developmental psychology in the late 19th century when the child's body and its physiological and psychological maturation became an object and field of interest (Prout, 2005). Piaget's constructivist theory linked mind to body development, creating measurable stages on the path to adulthood, suggesting that knowledge is only ever progressively gained (Bailey, 2024). Children become, therefore, "objects of concern, help and intervention and minimized in the capacity for dealing with their own problems" (Prout, 2005, p. 107). Our reimagining of HHH pedagogy thinks beyond the nature/culture binary and its humanist framing by viewing nature as inextricable from the-more-than-human-world, and as an enabling approach for working with children when thinking of them as capable of living in times of uncertainty (Kirby and Webb, 2023).

Reconceptualising HHH

Turning to indigenous cosmologies, Wall-Kimmerer, an academic who draws heavily on her first nation American heritage, explains that "we understand a thing only when we understand it with all four aspects of our being: mind, body, emotion, and spirit" (2013, p. 47), and contrasts this with her training as a scientist which foregrounds the 'head.' While recognising the benefits of the scientific method for humanity, Wall-Kimmerer (2013) suggests that it cannot tell the whole story, as it views knowledge creation from a position of separation. This notion of separateness is reflected in most curriculum frameworks which, even for the youngest children, tend to divide learning into discrete subjects or areas of learning. In contrast, Wall-Kimmerer (2013, p. 42) invites us to take a more holistic view of knowledge creation, "to see relationships, to seek the threads that connect the world, [and] to join instead of divide."

Similarly, writing from an animist perspective, drawn from the Korean tradition, Bai (2015) views the current western world view of the separateness of things, as having ethical consequences.

> If the earth is (seen as) mostly made up of insentient matter, and the biosphere is (seen as) made up mostly of mindless plants and 'dumb' animals, then it would follow that human beings with their self-concept of uniqueness and superiority would be inclined to dominate and exploit the earth and its inhabitants.

Bai (2015) suggests that we are not isolated individuals but enmeshed with the earth and biosphere, our existence dependent on a complex web of interconnections. This may seem a mere abstract idea expounded in an academic text, but it can be illustrated by a simple example; the act of tending a fire. The lighter of the fire adds twigs or some other fuel and the flames respond, perhaps burning more fiercely. The person tending the fire reacts to the fire's response. This reflects Barad's (2007) notion of an intra-action, where two entities meet and affect one another. Here, of the two entities, one is positioned as animate, one as inanimate, but the intra-action is perceptible.

In applying our post-humanist conceptualisation of HHH to pedagogy we argue that educators should engage with CaYP so as to encourage learning through the domains of head, heart and hands, in ways that consider the more-than-human world an equal partner in the learning process, so as to encourage respect, care and responsibility, whether the partner be plant, animal, fire or rock.

Exploring the concept of nature connection

The development of industrial societies has led to a decline in CaYP's exposure to and experience of the natural world as more people live in urban areas and spend more time indoors than ever before (Kesebir and Kesebir, 2017). Commentators argue that 'nature connection' is a psychological and physiological need, and consequently, there is a link between nature connection and happiness (Capaldi et al., 2014; Sellars and Imig, 2021, p. 1155). Indeed, Louv (2010) coined the term, 'nature deficit disorder' to reflect this. However, we caution that framings of nature as missing from children's lives suggest that children require exposure to nature and fail to take into account that nature is not unquestionably good (Bailey, 2020; forthcoming). Therefore, ideas and pedagogies for connecting children to nature need to be considered in ways that respect children's holistic needs including their physical, physiological and psychological welfare and the nature they will be experiencing. As we suggested earlier, pedagogy grounded in sensitivity and responsiveness to what is valuable and meaningful to CaYP in the present should not be sacrificed for an idea of what they should become (Moss and Cameron, 2020).

As biodiversity continues to decline, the wellbeing of the more-than-human world suffers. Hand et al. (2016) observe that children's lack of connection with nature leads to their ignorance and indifference to biodiversity, and a lack of motivation to respect and care for the world. This may be compounded by children experiencing a fear of nature. Ahi and Atasoyp (2019, p. 292), suggest that "For children to establish close relationships with nature, they need to enjoy nature, develop a sense of empathy towards other creatures, and learn that nature and humans are connected." However, for some children, for example those whose families are engaged in hazardous work on the land, and in climates where change is extreme, nature is far from benign, and is experienced in ways that are not beneficial for their physical and mental wellbeing. Therefore, fear is warranted in such cases, and respect for nature should be understood differently, depending upon the cultural construction of nature (Bailey, 2020; forthcoming; Guha, 2006).

Many CaYP live in environments which are nature-denuded and impoverished, with the UK being one of the most nature-depleted countries in the world (Reed, 2024). In her research in the USA, Wall -Kimmerer found that even her students who had chosen to study environmental science were unable to identify any positive relationships between humans and nature. She noted that "As the land becomes impoverished so too does the scope of their vision" (Wall-Kimmerer, 2013, p. 6). This exemplifies the failure of the education systems of the industrialised world to develop individuals and communities who can be advocates and activists for a sustainable future. This is perhaps concerning when children have become subject to a global 'responsibilisation' discourse (Walker, 2017). In the sense that governments and wider society, view children as those most affected by climate change, and, therefore, those with the most interest in, and capability to, bringing about sustainable attitudes and behaviours, irrespective of their contextual circumstances (Phoenix et al., 2017; Walker, 2017). In this regard, children's education on the issues surrounding sustainability is seen as playing a key role in positive change, as highlighted within the Sustainable Development Goals (un.org, n.d.). Yet children's agency to act, and to participate in any discourse, is always dependent on their relationships with those adults that take care of them (Walker, 2017). This challenges the degree of agency CaYP have to bring about change to achieve sustainability (Walker, 2017). As education is key to achieving change, any pedagogical approach supporting and encouraging children's agentic potential is key.

Taylor (2013, p. 50) suggests that notions of nature as a benign educator, and Rousseau's legacies of 'nature's child' endow nature with mythical and magical powers, and bequeath children with an intrinsically special relationship with nature. Thus, children's 're-connection' with nature becomes un-questionable. Speaking anecdotally from Early Years Practitioner (EYP) experience, enabling contact with 'nature' in UK Early Childhood Education and Care (ECEC), is frequently challenging and may be very limited in urban contexts. In addition, sustainability may be interpreted differently depending on the knowledge, beliefs and values of individual practitioners. When thinking about 'connecting' children with nature and the promotion of a post-human approach to a HHH pedagogy, we are interested in how this might be done in ways that respect nature, as part of a more-than-human-world that includes humans in it. We are also concerned with how we as pedagogues, encourage respect, an ethic of care, and responsibility, without assuming that contact with 'nature' will do the work for us. Below we consider an example of this.

A case study: towards a post-human HHH pedagogy

The following scenario took place during Kathleen's doctoral research, which adopted Barad's (2007) notion of intra-action – the moment when two entities meet and affect one another during a one-off art-making session in a wild park area (with thirty-four young children aged eighteen-months to nine-years-old). The research was designed around the observation that young children construct their ideas on nature through their relations with others, including those inhabiting what Taylor and Guigini (2012) refer to as the more-than-human-world. Kathleen narrates the following story of George aged six and the worms in the present tense, to recognise the way in which intra-action is always taking place; bringing the past and future into the present (Barad, 2007). Pseudonyms have been used to preserve individuals' anonymity. (See Bailey, 2023, for a more detailed account of this story).

George and the worms

George: *Wormy, wormy, wormy, wormy, wormy!*
Daisy: *Kathleen, George found a wormy*
Me: *I hope he's going to take care of it, let's go and check*
Daisy: *He's in the mud*

When I re-experience these words, I am struck by the way in which Daisy appeals to me and the way in which I respond, perhaps because I am an adult and moreover an EYP aware of children's needs, demands and actions but also maybe there is an agenda of familiarity where I am the person who might do something about an unfolding injustice. Therefore, I am propelled into acting.

Me: *What have you found George... George* (George is running away from me)
George: *He wants to go on my pic...*
Me: *Look at him!*
Milo: *Put wormy back!*
Me: *Wonderful*
Milo: *Don't hurt him!* (whine, high pitched)
George: *I'm not, he wants to go sticky on my picture*
Milo: *Don't stick him on there!* (whine, high pitched)
Me: *What about on the mud...Oh dear I hope that didn't hurt him... he dropped, fell a long way didn't he?*

Milo is also triggered into action, whilst my approach is that of a pedagogue seeking to enter into the worm play, so that I might guide George's actions, Milo's response is more emotionally driven. There is a sense that some form of transgression is about to occur that must be prevented. Perhaps Milo doubts my power to intervene. While my adult position endows me with authority, I am also an educator and carer, as such my 'gentle' EYP approach may not measure up to the real and present danger of dissection that this soft bodied invertebrate is in.

Some post-humanist scholars are critical of human anthropomorphising of the more-than-human world, yet in this instance I deploy the practice to draw attention to the worm's vulnerability as a living breathing sentient being. By agreeing with George's genderisation of the worm as 'he,' I seek to reinforce the notion that the worm is like George, a male. Perhaps we are hoping that as a male, George might identify with the worm and its chances of survival may be enhanced. George continues in his endeavour to stick the worm to his picture; in the process the worm falls to the ground.

George: *Where's wormy wormy?* [whining]
Me: *You don't want him to be hurt do you, he's very important to us*
George: *I'll put him in here... I'll make him a nice home* [loudly]
Me: *I think that's a good solution*

I wonder if Milo, as a child slightly older than George, represents a Vygotskian 'more knowledgeable other' in the 'zone of proximal development' (Corsaro, 2015). Milo's instructions to George were passionate and directive. I sense that had I expressed such orders they would have been ineffective or even inflammatory to the situation. Yet George's decision to make a home suggests care for the worm that seems to respond to and align with Milo's concern and values. For my part, I erred towards Sustained Shared Thinking (SST) (Siraj-Blatchford and Manni, 2008). Siraj-Blatchford and Manni (2008, p. 7) conceptualise SST as engaging in thinking or action with a child to prolong and improve a learning experience.

As an educator, when engaging in child/small creatures' intra-actions I seek to draw children's attention to the significance of our entangled human/more-than-human-world relationship. My view is underpinned by Lovelock's (2014) conceptualisation of the Earth as Gaia – a symbiotic self-regulating system. Every part of it, rocks, chemicals, oceans, atmosphere and organic species, including worms, play essential roles in keeping the planet habitable for "whatever is its biosphere" (Lovelock, 2014, p. 75). Framed in this way, the worm is important to us humans, without it the system becomes unbalanced. I am relieved when George decides to return the worm to his 'home' in the ground, as such (prompted by my knowledge of Bandura's notion of positive reinforcement (Nickerson, 2024), I praise George for finding a 'good solution.' Perhaps our combination of pedagogic guiding prompts George into exploring this notion of worm care as his play then evolves into role play with the worm. James and James (2011) argue that through play children make sense of their social worlds.

George: *Nice home wormy, wormy you got a nice home, wormy you got a nice home ahh... wormy where do you wanna come wormy* [with loud voice] *Ahh... aaa* [inaudible] *love you wormy* [with soft voice].

Yet, shortly after this caring role play, a worm death is reported by Isabelle and my adult authority is again called upon to dispense justice. On arriving at the scene, I ask to see the body, so that I might judge as to whether a death has occurred or if the worm might be saved by taking '*him to worm hospital*' where he might regenerate; a concept that I deploy regularly in my practice as an EYP. It transpires however, that the body cannot be found, therefore I pass judgement:

Me: *next time you find a worm … do you think you could … take a bit more care of him … because worms are very important to us, do you know why?*
George: *No*
Me: *Why do you think worms might be important to us … what do they do?*
George: *I don't know*
Me: *Well worms eat all the thing[s] on the ground, leaves and stuff to make mud and then we grow our food in the mud … do you like bread?*
George: *Yeah*
Me: *Do you like pizza?*
George: *Yeah*
Me: *Well if worms didn't do their job we wouldn't get pizza … or bread … or lots of other things, so that's why it's very important that we take care of worms. Do you think George next time you could take more care of the worm?*
George: *Yeah*

I see myself like (or as) a judge in my making a recommendation for how 'the accused' might improve. In doing so I cast the 'victim' as a worker in an economy of food production. Food and its production are a key area of concern in governmental concerns for sustainability (OEP, 2024). As I am aware that George is a fan of pizza, I appeal to his food preferences to portray the worm as indispensable, and a subject worthy of protecting. Worms work, within this food chain framing, is fundamental to human survival, therefore I link George's (human) welfare to that of the worm, conceptualising worm care as a practice congruent to the preservation of a certain kind of nature and hence, sustainability (Bailey, 2023).

As the session proceeds the children, with whom I have a worked with in the past and developed a trusting relationship with, continue to monitor and police George, calling on me, on one hand to administer justice, and on the other making me accountable for the worms' survival. Arguably, my post humanist approach has played a part in shaping the values of the children that now demand I enact the justice I advocate. This also leads me to ponder on George and the way in which he responds to our monitoring and policing. Davies (2014, p. 737) describes social exclusion anxiety as an individual's psychological engagement with emotional stresses to be 'included,' and hence the very notion of being social, as social inclusion is vital to one's ongoing survival as part of a group (Christians, 2018). I wonder if George's worm play is related to desires to be like these other children he plays with in Forest School, that care for small species, and that are praised by me for conforming to my post-humanist ideology.

In the recounting of George and the worms, I recognise how my pedagogical approach is informed by my ECEC practice and experience, my post-humanist values, and a desire to enable children to think of themselves, and their ongoing survival, as inextricable from that of the more-than-human world. Perhaps the way in which I am accustomed to talking about and with, small 'companion species,' has affected the way in which these children who know me as a pedagogue have come to extend care to animals such as worms. During the research the children seemed to draw on the existence of a moral code within their small social group that valued the life of worms (and other small species) that they regularly come into contact with when they play outdoors. I think of this as an enactment of 'interspecies justice' for which the children refer to their knowledge of (western) humanist justice systems to shape and enact their own form of justice. Positioning the worm as the 'victim,' and George as the 'individual that must be brought into line with the status quo,' the children took up positions of 'police and jury,' while bestowing on me the role of 'judge,' that must ensure justice is enacted. The analysis of this case study highlights how children are capable of engaging with "notions of right and wrong, and collective political activism [...and] compassion and responsibility" for more-than-human others (Bailey, 2023, p. 83), congruent with an 'ethic of care' and a HHH pedagogy. Moreover, it suggests that these children are far from accepted discourses which frame them as vulnerable and weak and disconnected from nature.

Conclusion

We began this chapter by suggesting that there was a strong case for overhauling the dominant discourses in education. We reflected upon the entreaty by Gallagher and Stobbs (2023, p. 2) to propose "ecologies as yet untried" and to imagine "human natures as yet unseen" and in response suggested that we recognise ideas and values meaningful to us and to CaYP in the present, while rediscovering and drawing upon knowledge and wisdom easily obscured by dominant discourses. Doing so, offers the possibility for a symbiotic more-than-human world view where humans have a role to play in maintaining a sustainable planet for all species. Therefore, we propose embracing Puig de la Bellacasa's (2017) conceptualisation of a more-than-human world as a possibility for realising a holistic pedagogy that recognises CaYP as potential agents of change and invites them to identify as participants in an ethic of care.

We are, in one sense, all activists in that our actions influence and are influenced by the world. An ethic of care acts as an antidote to the disempowering effect of disconnection. It encourages us to recognise the interconnectedness of all things and thereby expand our 'circle of ethics,' (Singer, 2011) to embrace a universal ethic of care. This represents to us what holistic pedagogy and the reimaging of HHH means in the age of the climate and biodiversity crises.

References

Ahi, B. and Atasoy, V. (2019) 'A phenomenographic investigation into preschool children's relationships with nature through drawings', *International Research in Geographical and Environmental Education*, 28(4), pp. 281–295. Available at: https://doi.org/10.1080/10382046.2019.1649248.

Bai, H. (2015) 'Peace with the earth: Animism and contemplative ways', *Cultural Studies of Science Education*, 10(1), pp. 135–147. Available at: https://doi.org/10.1007/s11422-013-9501-z.

Bailey, K.T. (2020) *Nature as good an ECEC product and practice.* Available at: https://circyatsussex.wordpress.com/2020/01/10/nature-as-good-an-ecec-product-and-practice/

Bailey K.T. (2023) 'Nurturing nature with(in) Children; or George killed the worm', in Solvason, C. and Webb, R. (eds) *Exploring and celebrating the early childhood practitioner: An interrogation of pedagogy, professionalism and practice.* London: Routledge, pp. 77–86.

Bailey, K.T. (2024) 'Listening "with" children "to" the more-than-human world: A diffractive analysis of the child and the challenge of sustainability', *Health Education Journal* [Preprint]. Available at: https://doi.org/10.1177/00178969241272865.

Barad, K. (2007) *Meeting the universe halfway: Quantum physics and the entanglement of matter and meaning.* Durham, NC: Duke University Press.

Capaldi, C.A., Dopko, R.L. and Zelenski, J.M. (2014) The relationship between nature connectedness and happiness: A meta-analysis. *Frontiers in Psychology*, *5*, p. 976. https://doi.org/10.3389/fpsyg.2014.00976.

Christians, C. G. (2018) 'Ethics and politics in qualitative research', in Denzin, N.K. and Lincoln, Y.S. (eds) *The Sage handbook of qualitative research.* Fifth edition. Los Angeles: Sage, pp. 66–82.

Corsaro, W.A. (2015) *The sociology of childhood.* Fourth edition. Los Angeles: Sage.

Davies, B. (2014) 'Reading anger in early childhood intra-actions: A diffractive analysis', *Qualitative Inquiry*, 20(6), pp. 734–741. Available at: https://doi.org/10.1177/1077800414530256.

Gallagher, S. and Stobbs, N. (2023) 'Creating hope in dystopia: Utopia as Method as social pedagogy in early childhood studies', *International Journal of Social Pedagogy*, 12(1), pp. 1–12. Available at: https://doi.org/10.14324/111.444.ijsp.2023.v12.x.005.

Guha, R. (2006) *How much should a person consume?* London: University of California Press, co-published with Permanent Black, Delhi, India.

Hand, K.L. et al. (2016) 'The importance of urban gardens in supporting children's biophilia', *Proceedings of the National Academy of Sciences*, 114(2), pp. 274–279. Available at: https://doi.org/10.1073/pnas.1609588114.

James, A. and James, A.L. (2011) *Key concepts in childhood studies.* London: Sage.

Kesebir, S. and Kesebir, P. (2017) 'A growing disconnection from nature is evident in cultural products', *Perspectives on Psychological Science*, 12(2), pp. 258–269. Available at: https://doi.org/10.1177/1745691616662473.

Kimmerer, R.W. (2013) *Braiding Sweetgrass: Indigenous Wisdom, Scientific Knowledge and the Teachings of Plants*. Minneapolis, Minnesota: Milkweed Editions.

Kirby, P. and Webb, R. (2023) 'Conceptualising uncertainty and the role of the teacher for a politics of climate change within and beyond the institution of the school', *Educational Review (Birmingham)*, 75(1), pp. 134–152. Available at: https://doi.org/10.1080/00131911.2021.1933392.

Latour B (2017) *Facing Gaia: eight lectures on the new climatic regime*. Cambridge: Polity Press.

Louv, R. (2010) *Last child in the woods: Saving our children from nature-deficit disorder*. London: Atlantic books.

Lovelock, J. (2014) *A rough guide to the future*. London: Allen Lane.

Moss, P. and Cameron, C. (2020) 'From "the state we're in" to "what do we want for our children"', in Cameron, C. and Moss, P. (eds) *Transforming early childhood in England: Towards a democratic education*. London: UCL Press, pp. 220–237.

Nickerson, C. (2024) *Albert Bandura's social cognitive theory. Albert Bandura's social cognitive theory* . Available at: https://www.simplypsychology.org/social-cognitive-theory.html (accessed 12 August 2024).

Office for Environmental Protection (2024) Progress in improving the natural environment in Engalnd 2022/2023. Available at: https://www.theoep.org.uk/report/progress-improving-natural-environment-england-2022-2023.

Phoenix, A. et al. (2017) *Environment in the lives of children and families: Perspectives from INDIA and the UK*. 1st edition. Bristol: Policy Press.

Prout, A. (2005) *The future of childhood: towards the interdisciplinary study of children*. London: RoutledgeFalmer.

Puig de la Bellacasa, M. (2011) 'Matters of care in technoscience: Assembling neglected things', *Social Studies of Science*, 41(1), pp. 85–106. Available at: https://doi.org/10.1177/0306312710380301.

Puig de la Bellacasa, M. (2017) *Matters of care: Speculative ethics in more than human worlds*. Minneapolis: University of Minnesota Press.

Reed, S. (2024) *Saving Nature*, UK Parliament. Available at: Written statements - Written questions, answers and statements - UK Parliament (accessed 15 August 2024).

Sellars, M. and Imig, D. (2021) 'Pestalozzi and pedagogies of love: Pathways to educational reform', *Early Child Development and Care*, 191(7–8), pp. 152–1163. Available at: https://doi.org/10.1080/03004430.2020.1845667.

Singer, P. (2011) *The expanding circle*. Princeton: Princeton University Press.

Siraj-Blatchford, I. and Manni, L. (2008) '"Would you like to tidy up now?" An analysis of adult questioning in the English Foundation Stage', *Early Years (London, England)*, 28(1), pp. 5–22. Available at: https://doi.org/10.1080/09575140701842213.

Soëtard, M. (1994) 'Johann Heinrich Pestalozzi', *Prospects*, 24(1–2), pp. 297–310. Available at: https://doi.org/10.1007/bf02199022.

Stein, Z. (2019) *Education is the metacrisis: Why it's time to see planetary crises as a species-wide learning opportunity*. Perspectiva. Available at: https://systems-souls-society.com/wp-content/uploads/2022/01/Education-is-the-metacrisis.pdf (accessed 2 January 2024).

Taylor, A. (2013) *Reconfiguring the natures of childhood.* Milton Park, Abingdon, Oxon: Routledge.

Taylor, A. and Giugni, M. (2012) 'Common worlds: Reconceptualising inclusion in early childhood communities', *Contemporary Issues in Early Childhood*, 13(2), pp. 108–119. Available at: https://doi.org/10.2304/ciec.2012.13.2.108.

Trodd, L. (2016) *The early years handbook for students and practitioners: an essential guide for the foundation degree and levels 4 and 5.* Abingdon, Oxon: Routledge, Taylor and Francis Group.

Tronto, J.C. (1993) *Maral boudaries : a political argument for an ethic care.* New York: Routledge.

Ulmer, J.B. (2017) 'Posthumanism as research methodology: Inquiry in the Anthropocene'. *International Journal of Qualitative Studies in Education*, 30(9), pp. 832–848.

Urry, J. (2000) *Consuming places.* London: Routledge.

Walker, C. (2017) 'Tomorrow's leaders and today's agents of change? Children, sustainability education and environmental governance', *Children & Society*, 31(1), pp. 72–83. Available at: https://doi.org/10.1111/chso.12192.

5

ENGAGING WITH UNCERTAINTY

A practical philosophical approach

Rebecca Webb and Perpetua Kirby

Introduction

This chapter explores a practical philosophical approach to engaging with sustainability education underpinned by democratic values. This means offering students and educators the opportunity to engage curricular knowledge and what lies beyond it, including the opportunity to be open to the uncertainty of what is encountered, so that they might allow themselves to be changed by what new things emerge. The significant democratic invitation is to be uncertain together, where neither the educator nor the student knows (and cannot know) how the student (and teacher) will respond in advance. This approach orientates the educational encounter towards the agency of the student to take up their equality and freedom (Rancière, 1991; Kirby and Webb, 2021), including to work out how to mitigate and adapt to sustainability challenges within the particularities of their own lives. We showcase a vignette of an adult and young child exploring the natural world together, where both have the opportunity to work out a response to the encounter. Aligning with the work of the educational philosopher, Gert Biesta (2021), this chapter explores the role of education in encouraging students to question what it is they desire and whether this might lead to damaging the planet, other humans, nonhuman groups, and themselves. This asks them to work out what else they might do in response to sustainability challenges beyond simply doing whatever it is that they desire, which includes both what they feel they want and what they might have been told to want.

First, we examine why practical philosophy matters educationally and why it needs to be put to work specifically to engage with sustainability issues so that we might each be called to act. We then go on to explore three models

DOI: 10.4324/9781003506850-8

of education that we have developed: the 'Plane,' 'Train' and 'Boat.' In our research and work with teachers and students, we use these models to help us become more aware of the shifts that we are making, moment by moment in our teaching, and to suggest why. The first two models are recognisable in education; both are commonly informed primarily by psychology and sociology. We suggest the value of a practical philosophy for all three is that the focus is on students' existing equality rather than simply a staged development to adulthood, in order that they might respond in ways that are unique to themselves. We apply these models using the example of an adult and their charge, and demonstrate the shifts made between all three in a short interaction, as well as the virtuosity of the adult educator in making judgements about when and how to shift between them. In the vignette we have selected, we highlight a young child's equality within each model explored. We take an early years example to demonstrate that a practical philosophical approach is appropriate for all students, regardless of age or stage, whether in schools, nursery, special schools, higher education or informal education. The vignette is about an adult in a park with a young child in their care. By virtue of 'studying' a slug, the child also happens to be a student.

Practical philosophy for sustainability education

Practical philosophy is about democratic, embodied and affective educational practices. Applied to the 21st century, dominated by a global climate and biodiversity crisis, it demands that we attend to the existential, to consider how we might live our everyday lives on a "damaged planet" (Haraway, 2016). The application of practical philosophy enables us to apply what we know to advance solutions to this crisis. How to live one's life well has been a central concern of philosophers (and others) for millennia. After the second world war, there was a particular emphasis on how to avoid global conflict and genocide. Hannah Arendt (Arendt, 1963) observed the trial of Adolf Eichmann, the Nazi bureaucrat responsible for the mass transportation of Jews across Europe. His legal defence was that he dutifully did as he was told. Arendt concluded that the banality of such evil comes from a failure to think and to engage with reality: an acceptance that had become terrifyingly normal, with insufficient challenge meaning evil could spread across the world like a fungus. Arendt (1963, p. 125) commented that "[Such] remoteness from reality and thoughtlessness can wreak more havoc than all the evil instincts taken together." Donna Haraway (2016), a contemporary philosopher, who picked up Arendt's work and applied it to the Anthropocene, has argued that a failure to think hampers an ability to respond. She captures this thus:

> In the surrender of thinking lay the "banality of evil" of the particular sort that could make the disaster of the Anthropocene, with its ramped-up

> genocides and speciescides, come true. This outcome is still at stake; think we must; we must think! ... [Eichmann] could not cultivate response-ability, could not make present to itself what it is doing, could not live in consequences or with consequence... The world does not matter in ordinary thoughtlessness.
>
> *(Haraway, 2016, p. 36)*

Within a practical philosophical approach, we can ask what her rallying call to think might mean for schools, other educational institutions, and educators. Biesta (2017) takes up this challenge to assert that there should always be the "question of direction" for education to enable the asking of "what it is *for*?" (Biesta, 2017, p. 15). Drawing on the work of Arendt, he frames what he calls a "world-centred" education. This is an education that does not centre on just the student or the curriculum but, instead, works to arouse the desire of students to exist in and with the world as a 'subject,' rather than an 'object.' The latter is driven by internal passions and desires and/or by external forces such as the desire and pressure to consume beyond what is sustainable for the planet. A world-centred education requires the 'I' always be relational to what is other in the world beyond the ego of the self. Indeed, the 'I' comes into being through encountering the other and taking up what this might demand, including when this is uncomfortable, in order to be a different sort of 'I.' This is the 'I' that is able to respond to climate change and biodiversity loss, even if when what is asked is contrary to an agreed status-quo. The role of the educator then becomes to encourage the student to question not what they might want from the world, but what is it that the world is asking of them. A key demand placed on the student, in this approach, is therefore to interrogate more closely the internal and external forces driving what they want:

> [This] always raises the question whether what one desires ... is what one should desire, in light of living one's life well, with others, on a planet with limited capacity for fulfilling all our desires.
>
> *(Biesta, 2022, p. 35)*

This practical philosophical approach is not located within psychological or sociological educational theory. Psychology assumes an intrinsic fixity of the 'I,' and sociology an identity as an 'I' (i.e. gender, race, class, and so on). Working solely within a psychological framing means that something internal has to be overcome in order for the 'I' to be restored so that one might be able to act; only working in a sociological framing means that something external has to be overcome in order to act. Both assume that one must overcome limitations to the formation of the self and/or to the social structures that reproduce injustices and differential opportunities and outcomes.

To consider what a practical philosophical approach brings beyond a socio-psychological framing, we can use the example of ecoanxiety. A practical philosophic approach challenges ecoanxiety as something psychologically discrete from education. Rather the framing assumes that it is through connecting to the world that one takes up one's freedom to come to a position and to act .This is not to deny the multiplicity of psycho-social feelings and concerns of the individual student and the external restrictions that they may encounter; rather it demands that it is through the staying with difficulty (without causing harm to oneself or others) that action can be enabled. A relationship with the reality of the world extends beyond the immediacy of one's own sense of self and one's desires, in order that we might respond to what the world demands of us. This is where the 'I' of the self becomes extended – to enable it to 'face out' – to become connected to the world, rather than only fixated on the psycho-social interiority of the self.

Living one's life this way, says Biesta (2022), is what it means to be 'grown-up.' Ultimately, neither educators nor students can sidestep the reality of sustainability issues to be a grown-up, regardless of age or stage. This is difficult and often painful to do; equally, it can be deeply satisfying when supported by others: educators, families, or communities. A practical philosophical approach opens the possibility to chart a way forward in which everyone takes on a role to address the challenges.

Examining models of education through a philosophical lens

To help us think with practical philosophy for education, including how we, as teachers, shift our practice in ways that remain educational, we outline three models that we have devised (Webb and Kirby, 2019). These aim to capture different purposes and practices of education that enable us to think about the role of education for transforming existing ways of knowing, doing and being.

Of the three models, the first two, the 'Plane' and the 'Train,' are about learning what is already known and valued, including what is set out in any prescribed curriculum. For the purposes of sustainability education, the assumption is that knowing things means a student will act in socially desired ways, either now or in the future, to ensure a greener world. The Plane emphasises the fastest and most efficient route to the learning destination, in which students are expected to sit facing forward ready to listen. It emphasises transmission from educator to student and being instructed in a 'right' answer and socialised into correct way of doing things. Teachers share knowledge, check comprehension, and ensure students' compliance. The Train also aims to reach the same destination but takes a slower more circuitous route. Students might work in groups and be actively engaged (rather than told) in discovering elements of the curriculum that the teacher has planned. The emphasis here is on play and experimentation to arrive at the learning destination. Both models are commonly dominated by psycho-social ways of ensuring the most effective and efficient route to reach equitable learning outcomes. A practical philosophical lens emphasises how in the Plane and the Train, students might assert their democratic equality through demonstrating their capacity to come to know, and to communicate what it is that they know and can do. Students are actively inserted into social orders – pre-existing ways of thinking, doing and being (Bingham and Biesta, 2010). They might, for example, take up the expectation to engage in 'green' practices, such as recycling. In other words, they actively demonstrate their own role in their learning and their socialisation, and through doing so they assert their equality to be included in the social order.

The third model is represented by a seafaring 'Boat,' that sets out without a clear destination. This is about engaging students with uncertainty, where there is no expectation that either they or the educator is required to know a clear answer in advance, but for each to share what it is that is already known. The focus is on something where the answer cannot be known or is contingent, allowing for something new to emerge individually and/or collectively through the process of engagement. This might include a response to a prompt that is written or a real life problem, where there may be existing commonly agreed assumptions about what is the 'answer,' but equally space for each person to come to their own response. Examples could include interpreting 'nature' poems replete with metaphor (such as The Windhover by Gerald Manley Hopkins); decisions about where to place wind turbines; dilemmas of mining rare earth minerals for electrical batteries from communities with fragile ecosystems and threatened local livelihoods.

Teaching methods can be varied but creative-deliberative approaches can help to amplify the uncertainties, such as are used for philosophy circles of inquiry or citizen assemblies, or a personal response through the arts. Here, there is less emphasis on what is being learned, and more on how an engagement with the world might transform existing ways of thinking, being and doing. Students come to unique positions and decisions for action. The educator's role is important in requiring students to stay with difficulty, pay close attention, think deeply and verify positions. It does not allow students to share opinions and make claims that cannot be substantiated and verified, through discussion with others or by reference to other forms of knowledge sources.

Educationally, when in the Boat, students can and must be able to disrupt a social order. This is crucial for democracy. The Boat offers the possibility for students to take up their freedom in ways that are not pre-designated for them. They might challenge existing ways of doing things, such as school practices considered environmentally unsustainable. For example, demonstrating against the selling of plastic bottles from dispensing machines, or undermining the dominant human extractivism of nature through refusing meat-based school lunches. In these examples, students demonstrate their equality to transform that which has come before; they challenge the assumption that only as adults, once people have acquired certain knowledge and skills, might they be equal to communicate and act on their concerns.

Shifting between models: a vignette

We illustrate the above three models using an example of a young child who shows an interest in slugs. Through doing so we wish to evoke what it is that a practical philosophical lens can bring to sustainability education, where what is important is less what we know but what we might do with what we know, in order to mitigate and adapt to climate change and biodiversity loss. We use a personal vignette, in which one of the authors (Rebecca), spent time with a young charge, in a seemingly non-educational context which was an outing to a local park. We bring our own ethnographic sensibilities as researchers to both the telling and the revisiting of this park visit. In so doing, we evoke the memories of the specificity of the time and place, which includes the emotions, scenic and materiality. Through a shared narration we engage in a 'grown-up' reflexivity that holds each author to account as we revisit and interrogate the story in relation to the three models of education and the judgements made by Rebecca, the educator, at the time.

Working with young children amplifies both the opportunities and dexterity in the shifts between the three pedagogical models, to assert the benefits of a practical philosophical approach. In his book, *The Beautiful Risk of Education* (2013), Biesta explores the idea of the 'virtuoso' educator. They are those who never merely repeat what has happened in the past, but are always open to what a new situation requires and demands. Indeed, the practice is always about becoming an educationally wise person. This requires judgement rather than the following of recipes, to remain open to what is required at any given moment in relation to the student. We find this helpful when thinking about the role of educator shifting between the different models of education, as outlined above, as well as the demands on the educator to stay with the difficulty of being 'in the Boat' alongside the student. This requires the constant making of judgements about how to shift wisely, moment by moment, in response to what it is that students are doing.

The vignette that follows demonstrates that the philosophical approach is not about a particular educational institution or the child's age or development. By focusing on a non-institutional interaction, we draw attention to what is often colloquially represented by the idea that 'it takes a village to raise a child.' An education for sustainability action crosses the boundaries of the school (or other institution) walls to include those in the wider community, as it is inherently about how to live one's life well in the everyday. This is as true for young children as it is for older students and adults.

The story starts with Rebecca who is with Poppy, aged 25 months old, on an unseasonably warm early June day. They are in their local park, sitting on a rug under an old elm tree for shade. At the base of the trunk are multiple sprouting unfurling tender green shoots. Poppy finds a slug on a shoot, still moist from rain the night before. She signals to Rebecca to join her and to look at the slug that she has found. There is a dog sniffing around close by, that Poppy notices; she scoops up the slug into her small palm, saying 'keep slug safe, 'Ecca.' The slime means that the slug attaches to the palm of her hand, but with attention and caution she gently shakes it into her small plastic bucket. She then purposefully looks up and sets off away from the tree and the dog, to a more open space of grass, beckoning to Rebecca to follow. She bends down and carefully tips the slug out of the bucket onto the grass.

Being in the train

Looking back on this moment, Rebecca recognises that she was observing Poppy through an educator's gaze, to see what it was that Poppy might discover about slugs. Poppy already knew something about them, from being with Rebecca previously in her own garden, where she had helped to move slugs away from flowering plants to prevent them being eaten. Being in the park offers the opportunity to scaffold what Poppy knows and can do from one context to another. Poppy clearly knows that a slug is a living thing; that living things need care; and that we watch out for possible dangers that might harm ourselves and others. Poppy is wary of dogs and this one she meets is about the same size as her. She seems to sense a threat from the dog to the slug, but perhaps also to herself. She demonstrates her agency in doing something about this possible threat by moving herself, and the slug in the bucket, away from the dog, who is now sniffing around under the tree. Rebecca reflects on what she might have done when practicing as a teacher previously observing her young students. For example, she might have noticed that Poppy demonstrates motor skills through her ability to get the slug into the bucket, as well as a spatial awareness through moving away from the perceived threat. Operating in the train model, Rebecca is observing Poppy's learning. In other words, how Poppy is discovering things she is meant to come to know in the world; in an early years setting, Rebecca may well have checked these off on a list, even if only mentally. In practical philosophical terms, Poppy is inserting herself into the world of slugs, dogs, trees and humans, to which Rebecca responds, follows and attends minutely.

Being in the boat

Thinking about the boat in reflecting on this moment, we can consider what more is happening. Poppy is closer to the earth than Rebecca, both literally, being so small, and metaphorically: she is drawn to the slug and she is a master of attention to the minutiae of the more-than-human natural world in which the slug becomes the protagonist in this story. The dexterity of Poppy's body enables her to bend, move, shift pace, in a way that it is less easy for Rebecca. Poppy's inclining embodiment displaces the dominance of the uprightness and disconnection of the adult who is assumed to command the educative situation as the one who knows what it is that is worth paying attention to.

Poppy's engrossed involvement with the world draws Rebecca's own deep attention to the large plump sludge-brown slug. Even at the point of evoking the memory of the slug, Rebecca is revolted. Poppy on the other hand is clearly not repelled by it, including when its stickiness is hard to remove from her hand after she has tipped the slug onto the grass. In that moment, Rebecca has to consciously try to find something of beauty rather than being viscerally repulsed. This calls on her to adjust an initial reading of Poppy and the slug. By aping Poppy's embodiment, Rebecca too comes to know the slug in a different way. Rather than staying upright and being worried about getting slug slime on her, she must get down alongside Poppy. Confronting things in oneself, in others, and in the world that are different is clearly uncomfortable for Rebecca, both physically (as she bends) and intellectually, as she works out how to live alongside things that are slimily different. Poppy is also working out how to live alongside both slugs, dogs and adults. She takes up her freedom here to act in ways that are not pre-designated for her, in any curricula or by the expectations of Rebecca, and she does so through an openness to engage with difference. Both are core to a democratising practice. Poppy disrupts the social order through her wonderment of the slug, as well as her attempt to care for, rather than to dominate, kill or extract from nature for human purposes.

Taking the helm in the boat

Poppy's seemingly 'free play' in the park can take on a different configuration as either the Train or Boat depending on the situated reality between the educator and the child. Being in the Boat demands something of Rebecca, as an educator, that goes beyond observing and getting alongside Poppy where Poppy might do as she pleases. In the Boat, an educator pays attention to the students' actions and embodiments as reflecting their current knowledge and assumptions, observed through what they say and do. Rebecca, for example, knows that Poppy has learned to carefully move slugs, as pests, away from young plants. The park offers a different context, where Poppy transfers this knowledge to feeling a need to remove the slug carefully, but on this occasion to keep it safe from the dog. Staying in the Boat, Rebecca resists the urge to know more than Poppy about whether this is the 'right thing' to do or not;

she does not instruct her in what is best for the slug. What she does do is to pay greater attention to what it is that is driving Poppy's action and to consider, with Poppy, whether what she desires is indeed desirable for the slug and its connection with the wider habitat, as well as potentially for Poppy and Rebecca's sense of entitlement to know better than the slug what is best for it.

To enable Poppy to engage philosophically requires Rebecca to ask verification questions. This is more demanding intellectually and emotionally, both for Poppy and Rebecca, than simply affirming the value of Poppy's free play. Affirmations can easily slip into a gendered emphasis on kindness and care (often reinforced by women to girls) (McDowell Clark, 2022), as well as anthropomorphising the slug. Through our reflexive storytelling, Perpetua and Rebecca subsequently identify verification questions for Poppy that Rebecca might have asked at the time, had she more explicitly taken the helm in the boat. For example, she might have asked: Does the slug need help? How do you know? Is the slug safer here or by the tree? Rebecca is confident that Poppy would have comprehended this questioning, even if she did not have the verbal skills to offer full answers. Being in the Boat does not assume or need a verbal response from two-year-old Poppy. She might respond through action or through a shift in what she thinks and feels. Poppy might semi-address these questions in the moment, or else something might follow at a later point. She might reject the questions, and the thinking they require, altogether. We would not necessarily know. This, as Biesta suggests, is the beautiful risk of education.

The educator cannot control what happens in the Boat; they cannot assume they know the answer; nor should they signal a whiff of knowing the right answer, which will only close the space of democracy in this uncertain educational interaction. Educators need to remain curious, while also demanding deep thought from both students and themselves. Their role is to engage the student in verifying their assumptions and the validity of their actions. At this young age, it might just be modelling verification questions, such as 'I wonder if …,' in such a way that they hold out the possibility that more needs to be known, as well as the value of acquiring new knowledge. As Poppy gets more acquainted with philosophical questioning, she will be able to raise more sophisticated queries, such as: 'Do slugs fear dogs?'; 'Should humans kill animal pests?'; 'Should humans intervene in the wider natural world?'; 'Are slugs beautiful?' This process of verification is central to encountering climate change and biodiversity loss in democratic systems, where it is necessary to challenge assumptions about how to live in the world without further damaging other species and the planet, as well as ourselves.

When in the Boat, there is always the possibility to climb out and onto the Plane or Train, whether briefly or for a more sustained period. Philosophical questions lead students to a place where they need further knowledge, and the educator might be a source of new information and expertise. The educator is always attuned to the need to socialise students, particularly in relation to their safety, although they, too, can be wary of particularly socialising orders that might unnecessarily constrain. When in the park with Poppy, Rebecca repeatedly shifts to being in the Plane, when firmly telling Poppy to put her hat back on as the sun is hot. Rebecca might also have shared something of what she knows about slugs, and she might have stayed in the Plane on the return home, via a visit to the library, to find a book that includes relevant facts. The virtuosity of the educator comes about by embracing the uncertainty of knowing when and in what way to step in, so that they and their students might shift between the Plane, Train and Boat.

Conclusion

This chapter has explored the significance of the uncertain dynamic of engaging with sustainability education regardless of the age, stage, or context of the student. It has asserted that democratic education must provoke encounters that offer opportunities for students (and their educators) new ways of knowing, being and doing that could not have been anticipated in advance. This is an essential facet of educating for the dynamic, fast-paced and changing world of the 21st century where the urgency of the climate crisis and of biodiversity loss means that many responses and/or solutions are as-yet-unknown and, certainly, not yet enacted. Humans are required to pose difficult and compromising questions about how we might live democratically on a damaged planet, and the multifaceted decisions that are required of them individually, communally, and institutionally so to do.

We have drawn on philosophical works, especially that of Gert Biesta, to assert the importance of democratic education to offer both the student and the educator ways of taking a 'reality check,' utilising their agency and freedom to ask whether what it is they desire is, indeed, desirable, beyond established ideas of normative need. We have suggested that a practical philosophical approach assumes students' equality to insert themselves into established and existing social orders to allow for their thinking and questioning, regardless of whether the education model is that of the 'Plane,' 'Train' or 'Boat.' Using the vignette of the young child and 'her teacher,' we have illustrated how shifts between the Plane, Train and Boat might enable students to be transformed by educational encounters such that they can respond in unique and unexpected ways that cannot be foreseen or predicted by an educator.

Such unpredictability brings us to a close of this chapter, but also opens new and uncertain questions for the authors: in Poppy's encounter in the park (with her adult human, the verdant summer foliage, the slug, dog, and bucket), who (or indeed what) might lay claim to be transformational educator, and in what ways? The significance of the question for sustainability education lies in the possibility that the more-than-human may be the educator that draws the students' attention to what it is that the world might demand of them.

References

Arendt, H. (1963) *Eichmann in Jerusalem: A Report on the Banality of Evil*. London: Penguin.

Biesta, G. (2016) *The Beautiful Risk of Education*. London: Routledge.

Biesta, G. (2017) Educational Leadership for What? An Educational Examination. In D. Waite and I. Bogotch (Eds), *The Wiley International Handbook of Educational Leadership*. Chichester: John Wiley, pp. 15–27.

Biesta, G. (2022) *World-Centred Education: A View for the Present*. Abingdon: Routledge.

Bingham, C. and Biesta, G. (2010) *Jacques Rancière: Education, Truth, Emancipation*. London: Continuum.

Haraway, D. J. (2016) *Staying with the Trouble: Making Kin in the Chthulucene*. London: Duke University Press.

McDowell Clark, R. (2022) The Emergence of the Early Childhood Practitioner. In C. Solvason and R. Webb (Eds), *Exploring and Celebrating the Early Childhood Practitioner*. London: Routledge, pp. 7–8.

Rancière, J. (1991) *The Ignorant Schoolmaster: Five Lessons in Intellectual Emancipation*. Stanford, CA: Stanford University Press.

Webb, R. and Kirby, P. (2019) Modelling Transformative Education. *FORUM*, 61(1), pp. 89–103.

6

EMBEDDING SUSTAINABILITY IN THE EARLY YEARS CURRICULUM

The Curiosity Approach and its foundations in the work of education pioneers

Samantha Sutton-Tsang and Carla Solvason

Introduction

This chapter is based upon our research with an international school that has adopted the Curiosity Approach (CA). We use this research as a springboard to investigate how many of the approaches taken relate to concepts of sustainability. As we do this, we unpack the origins of many of the core approaches of the CA, relating them back to the ideologies of the principal educational pioneers. We dig a little deeper into the 'why' behind the 'what.' In exploring these approaches, we share some accessible ways of beginning to embed environmental awareness and responsibility into an early years setting and the curriculum. Finally, we consider what more is needed, to further develop activities that align with sustainability into a values-led approach, that is founded upon connectedness, with one another and with the planet.

As researchers (and teachers) based in England we were very excited to be given the opportunity to embark on a project with an international school in the United Arab Emirates (UAE) to explore how they had incorporated the Curiosity Approach (CA) within their setting, and how they embedded sustainability within that. This research provided an opportunity to work with early years practitioners from across the globe in a prosperous country with high living standards, which is identified as one of the lowest ranked countries for sustainability according to the Climate Change Performance Index (CCPI, 2024). Initial discussion with the study school indicated that there had been significant improvements in practice brought about through adopting the CA within the Early Years. Because of this we decided to carry out a case study of the implementation, adopting an appreciative inquiry approach (Cooperrider and Srivastva, 1987), to identify the perceived benefits.

DOI: 10.4324/9781003506850-9

Although the distance brought with it challenges, technology enabled us to observe videoed 'learning walks' through the setting, extensive documentary analysis and focus group discussions. The project enabled us to better understand the CA's application in practice, and how resources can be developed to provide a more sustainable learning environment that encourages children's guardianship of their planet.

All data was collected respectfully following full institutional ethical approval. Through working with the school, we discovered that the CA draws on a wide range of pedagogies, made up of the 'loose parts' of the conceptual jewels of previous pedagogical pioneers. The research offered the opportunity to revisit some of the strongholds of Early Childhood Education and Care (ECEC) thinking that have endured due to their embeddedness in the notion of the agentic child who is respectful and careful of the fascinating and fragile world that surrounds them. In this chapter we argue that the positive and hopeful concepts that the CA combines are not new, rather they are ideologies that have become buried in decades of varying socio-economic agendas. We suggest that now, in this period of the Anthropocene, it is a good time to rediscover some of the values that underpin these approaches and consider how they relate to more sustainable futures.

Founded in 2017, the Curiosity Approach was the brainchild of two UK-based early childhood practitioners, Stephanie Bennett and Lyndsey Hellyn. Bennett and Hellyn are successful businesswomen and qualified practitioners, with directorship of six nurseries across the Midlands of England. They recently featured on the BBC as the "nursery with no toys," were voted Top 10 Day Nursery in the West Midlands on daynurseries.co.uk, and have "multiple Ofsted rated 'Outstanding' sites" (The Curiosity Approach, 2024a). Such is the success of this package that it has made its way across the world. We were intrigued by how this approach could cross cultures to become installed within an international school in the Asian continent. We were also curious as to the relationship between this approach, with a focus upon natural resources, and wider concepts of sustainability.

The Curiosity Approach is a combination of existing theories of high-quality pedagogical practice, established over the last century, developed to be easily accessible and relevant to the 21st century. There is currently very little literature which explores this approach apart from the creators' own materials, hence the overlap between explanation and 'marketing' that can be seen in some of the references that we use. The combination of approaches that the CA promotes, speaks to a world in crisis, a world that has become painfully aware of its mortality and fragility. As Warden (2022, p. 94) advises, "One way that we can all impact positively on the survival of our species is through our teaching," and the CA offers some simple and easily accessible ways for practitioners to begin to embed environmental education within their teaching. They frame it in this way:

> Sustainable practices aim to strike a harmonious balance between ecological integrity, social justice, and economic prosperity, ensuring a world where both present and future generations can thrive. Early childhood education lays the foundation for lifelong learning, and it is crucial to instil sustainable values in children from an early age. By introducing sustainability concepts, we foster a sense of responsibility, empathy, and stewardship towards the environment and society.
>
> *The Curiosity Approach (2024b,* no page*)*

Below we explore some of the CA's proposed approaches, digging a little deeper to acknowledge from whence they originated, and the foundational pedagogical values that underpin them.

The autonomous child

The CA promotes environments which encourage children to be increasingly independent. For example, practitioners in our study setting mentioned making sure that equipment is available for children to access independently, or allowing children to use the toilet by themselves as a starting point; but as important as the context for children to become independent *actors* is the encouragement for children to become more independent *learners*. This is a first step towards empowering children to becoming "active global citizens, capable of making informed decisions that can positively impact the world" (The Curiosity Approach, 2024b).

The CA (2024a) suggests that practitioners present children with 'invitations to learning,' "magical pockets of learning" throughout the setting that "ignite... curious minds," with the ultimate goal to provoke ideas and encourage children to discover and investigate the resources and world around them. If we are to imagine a more hopeful future it is vital that children recognise themselves as agentic, and that they begin to recognise the role that *they* can play in creating more sustainable futures. Davis and Elliott (2014, *no page*) call for "empowered citizens, capable of both contributing to and creating change for sustainability." They call for children to develop a sense of social responsibility, something that has previously been embedded in historical approaches to the Early Childhood Education and Care (ECEC), but, perhaps, has gone relatively unnoticed.

Doull and Ogier (2024, p. 36) explain that children's empowerment is important, because it creates potential to develop "compassionate connection, convivial relationship and civic capacity." Similarly, Tē Whāriki (Ministry of Education, 2017, p. 6), New Zealand's Early Years curricula, presents a "vision that children are competent and confident learners and communicators, healthy in mind, body and spirit, secure in their sense of belonging and in the knowledge that they make a valued contribution to society." In England, the Early Years Foundation Stage (DfE, 2024 p. 9) provides a target that children "develop a positive sense of self" and "persist." This *persistence* is vital if children are to nurture a love of their planet within an environment that has long made decisions based upon extractivism and indifference. But the term lacks something of the agentive and holistic approaches to autonomous learning that were developed by some formidable early childhood pioneers over the past century.

The Montessori approach contends that children are "naturally motivated to learn" (Carnie, 2003, p. 71). When she opened her first school in Rome in 1907, Maria Montessori's view was that children are, by nature, "active learners" who need autonomy, and that the practitioners' role was to provide an appropriate learning environment for children to engage with (Isaacs, 2018, p. 1). Montessori argued that the child should be respected and trusted "to spontaneously select activities and engage in them, learning through problem solving and discovery" (Isaacs, 2018, p. 15). In other words, the child is agentic, and will find their own way without us directing them with carefully placed 'breadcrumbs.' Montessori (1964, p. 95) argued that instead of leading a child towards answers, we should "lead the child toward independence." For Montessori (1964, p. 87) the spirit of the individual child is key:

> Humanity shows itself in all its intellectual splendour during this tender age as the sun shows itself at the dawn, and the flower in the first unfolding of the petals; and we must respect religiously, reverently, these first indications of individuality.

Montessori education nurtures environmentally conscious individuals. These children are better placed to understand the interconnectedness of people and the environment, preparing them to contribute positively to sustainability efforts. Montessori also recognises that adults must also be prepared to learn with and from children, arguing that prescriptive, transmission approaches, to education do no more than maintain the existing status quo. And within our current context, that will mean no more than perpetuating the existing disconnectedness between children and their planet, and the ongoing environmental degradation.

If practitioners are to create a generation of thinkers who are ready to challenge, then they themselves must be prepared to be challenged. Although the practitioners that we spoke with shared that they had become far more reflective practitioners through their development of the CA, this reflection appeared to be instigated more by theories and less by the discoveries of the children that they worked with, suggesting that we, as educators, still have some distance to travel in this area.

Reggio Emilia's approach to nurturing children's autonomy is similar to Montessori's, but goes further into trusting children to develop their own knowledge, skills, and independence, recognising their potential as agents of change (Thornton and Brunton, 2014). Central to the Reggio Emilia approach is the concept of adult and child as co-explorers in unknown territories, where "Priority is given to research between adults and children as an everyday praxis, a necessary attitude for interpreting the complexity of the world, and a powerful instrument of renewal in education" (Reggio Emilia Approach, 2022, *no page*). This nurtures the constant potential for change and challenge, for alternative knowledges and alternative futures. Malaguzzi (1990), a key figure in the development of the Reggio Emilia approach, described the situation thus:

> We are faced with phenomena of extraordinary importance that for the first time make it possible for young people to distance themselves from a measure of reality they were perhaps running from, escaping, not helped by families, or by schools, or by culture, or by the measure of political attitudes. For the first time they can autonomously imagine for themselves what it could mean to have a world free of something, not yet free today or fully capable of taking the ethics of co-responsibility to the level of credibility of something great. These subjects do not hover or circle above education: these subjects are part of the themes, the sensibility and the restlessness, that must be inside schools ... if they are to be capable of listening to the serious things rolling down the road and not remain silent about the restlessness and questions children and students have inside them.

This is similar to Deleuze and Guattari's (1987) concept of the nomad. This rejects the concept of a single line of thought, or a single line of development, and asks pedagogues to bravely encourage children to be autonomous learners, thereby creating opportunities to forge paths that are previously untrodden, to consider, and to learn, in different ways. If we truly want to change the future of our planet, it will take more than just allowing children 'opportunities' to learn what we already know, we must give them the space to imagine a different future, one that is coloured by the hope of rejuvenation.

The enabling (and natural) environment

Home from home

A key aspect of the UAE international school's adoption of the CA was the holistic transformation of the physical and sensory environment of the school.

This involved the replacement of plastics and primary colours with materials that helped to create a more muted, comfortable environment, that would more closely resemble the child's home. This included lamps and framed pictures, plants and sofas. This was applied to the whole environment, not just the rooms where children 'learn,' but everywhere from the rocking chair in the entrance hall to the real cutlery and crockery used at mealtimes.

In Hygge and Danish pedagogical approaches, the welcome into the setting should be warm and inviting, especially important with their cold dark mornings. Pedagogues do this by lighting candles and baking bread so that families and children arrive to lovely smells and a homely environment (Williams-Siegfredsen, 2017, pp. 36–37). Although the cold is not an issue in the UAE, adopting the CA encouraged the practitioners in our study to consider the emotions evoked by their school environment, and to introduce calming music and images, and pleasant smells. Through this, children are encouraged to more wholly engage their senses with their environment.

In Reggio Emilia settings, the carefully planned school environment is described as 'the third teacher'; this care taken when planning the environment was reflected in the practitioner's comments, from the international school, when they mentioned Teaching Assistants taking responsibility for making a certain area "inviting," with soft furnishings and muted colours. This transformation of the setting, our respondents shared, had the effect of calming the children, as one practitioner explained:

> The settling in period at the beginning of the year was much more relaxed. Many of the children had never been to nursery or school before due to Covid, so the feeling that school was an extension of home helped to create ease and calm.

Nicol and Taplin (2012, p. 2) explain how a calm environment is also important within the Steiner approach, because "a secure and unhurried environment enables children to develop a range of skills, which provide a sound foundation for emotional, social and cognitive intelligence later." This aligns with Maslow's Hierarchy of Needs (1943) whereby basic needs must be met before children can go on to achieve. Emotional and physical security is a basic need; only calm minds can flourish. This calls into question how prepared for learning our children are in the perpetual state of high anxiety that we live in today.

Children in the early years need to "develop… their sense of self and their confidence to explore" and Doull and Ogier (2024, p. 38) explain that this is best done from a point of comfort, with "interactions between home, school and the wider community…foster[ing] a sense of belonging." Belonging and comfort go hand in hand, as do belonging and responsibility. If we want children to feel responsible for their environment, then they must feel that they belong within it.

Montessori was also a proponent for realistic and homely environments in settings, based upon the belief that this can provide opportunity for children "to take responsibility, to manage risk and to learn that resources and equipment need to be treated with respect" (Hellyn and Bennett, 2019, p. 67). Hellyn and Bennett (2019, p. 67) clarify why this is important, in that it enables children to "understand that actions have consequences" and to understand that they must "take ownership of their actions and choices."

In Reggio Emilia, objects are "enticing" to encourage handling, as well as "functional," displayed on open shelves for easy accessibility to children (Feez, 2010, p. 42). We need our future generations to take interest in, and develop responsibility for, the world around them. The UAE school does not shy away from objects such as breakable glasses, in the belief that not only does handling fragile objects make the children "more independent," but they also make them more careful within the setting. In Reggio Emilia small, breakable objects are found around the setting to encourage children to take more care (Feez, 2010, p. 43), based upon the belief that children live in a fragile world and must be aware that their actions might harm or heal. Just as the research setting was eager to share that they had less breakages since their CA journey than at any time previously, as practitioners we must be more willing to give children responsibility if they are to learn to be responsible.

Natural and sustainable resources

Within the CA, in order to better reflect the world around them, natural colour tones are used on walls, display boards and furnishings. Our study school felt that nature was 'celebrated' throughout their setting. Maximising daylight, natural resources and care for plants all have the potential to draw children closer to nature, and are not new within ECEC theory. Baloch et al. (2020) researched the effects of daylight on school performance and found there was a significant relationship between light-related variables and performance in school. Thornton and Brunton (2015) describe the Reggio Emilia environment, with its windows and natural light, its free flow spaces for sharing and connecting with others, its subtle colour palette and its multisensory approach, as lessening the divide between the child and nature, something which is much needed in an environment where children are frequently surrounded by man-made materials and technology.

One practitioner in the study school explained that the CA was: "kinder to the planet because you can use materials that are readily available from the environment." They found uses for the materials that many throw away. Such resources abound in the UK, through charity shops, recycling plants and car boot sales for example, but in the UAE it is more of a challenge. This practitioner explained:

> Finding seasonal resources is difficult here. The Dubai climate rarely changes and the simplest of materials are not available to collect. They need to either be bought online or donated by parents or staff...Charity shops, car boot sales etc do not exist here, rendering a "good find" almost impossible. We rely heavily on the donations of parents/ carers... Resources have been curated from local markets, souks and junk shops because there are no charity shops in UAE...Staff have used family contacts, own scavenging and local contacts to resource the setting.

Within the UAE setting, construction materials are evident in each classroom. They refer to '*intelligent*' recycled resources such as containers, tubes, trays and fruit packaging and how these can be gathered and sorted. The practitioners also described how these resources were used with '*more respect*' by the children, if there was a place for them, and they are not just cast aside. The UAE setting shared that "*All team members are encouraged to recycle and reuse*" but the CA literature goes no further to explain why the use of sustainable resources is so important, so again we need to turn to other theorists for a deeper understanding of these concepts.

Thornton and Brunton (2014, p. 51) explain how the concepts of curiosity and creativity are central to Reggio Emilia, and how these are evidenced in the resources found within settings, from manmade resources such as glass bottles and beads to natural materials such as driftwood, stones, shells, cones, leaves, seed pods, feathers. Such resources do not replace the more traditional equipment that we would expect to see in ECEC settings, but they accompany them. When they are used, there is variance, in tools, textures, shapes and sizes to encourage variety as opposed to conformity. Within the Montessori approach natural resources go beyond being "aesthetically pleasing" (Isaacs, 2018, p. 83) and toward making nature visible. Breaking down the barriers between child and their planet that our manufactured world has created.

Steiner philosophy sees education "as a process of gradual awakening or unfolding of the self...that we develop our identity and the inner strength to deal with life's challenge," therefore by adopting a holistic approach, we can help children to make sense of the world as they grow and become part of society (Carnie, 2003, p. 42). Steiner's approach to education is that it is practical (developing skills in cookery, clothes making, gardening, wood and metal work), artistic (through art, sculpture, music, and drama) and will develop the intellect (to form opinions, social skills and spiritual values), to understand oneself and their place in the world (Carnie, 2003). This, Wilkinson (1993, p. 37) argues, is very different from preparing children for a world of work, as is the current educational policy preoccupation in England. Wilkinson (1993) explains:

> Contrary to current ideas it is not the task of education to guide a child into a profession. Given an education which embraces his whole being and enables him to find his right path in life, there is no need to choose a career. It will come naturally. In our present state of evolution, that is, in the midst of our materialistic civilisation, the essential thing for the individual is not to acquire more factual knowledge, even in the technical field, as some would advocate, but to find himself.

The CA promotes the upcycling and use of naturally foraged materials to resource the early years setting, avoiding the high costs of purchasing new items and buying into consumerism, which has significantly increased since COVID-19 (ONS, 2022; Shaw et al., 2022). Friedman (2008, cited in Warden, 2022, p. 6) argues that consumerism has played a key role in climate change, exhausting resources and causing "irreversible damage to ecosystems." These, it would seem, are all valid reasons for embedding sustainability within education, and justifications that remain unexplored in the CA literature.

Multi-sensory

The multi-sensory nature of the learning environment was apparent throughout the data that we collected from our study setting. The international school practitioners aimed for a "Hygge ethos including aromas, textures, lighting, sounds, the use of display boards, cozy areas for recharging." One practitioner shared how "*In the morning [the children] choose the oil for the diffusers and the music for the day.*" When discussing the playdough station that they had created, the practitioners discussed how the children were encouraged to mix it with ingredients that were easily accessible to them, such as dried herbs, flowers and fruits" adding a multi-sensory dimension to the activity. Once again, the CA reasoning behind this goes unmentioned, and we need to turn to EC pioneers to further unpack the reasoning behind this.

Proponents of Steiner Education stress that "learning flourishes in a calm, peaceful, predictable, familiar and unhurried environment that recognises the child's *sensory sensitivities*" (Hellyn and Bennett, 2019, p. 48, *our emphasis*). Steiner's philosophy for early learning was that through the use of all senses, interacting with, and actively engaging with the world around them, children can then develop the 'readiness' to engage with formal learning. Likewise, Schaefer and Kaduson (2006) refer to the 'physical connection' that a tactile material like playdough creates for children. Nicol and Taplin (2012, p. 14) explain how Steiner's principles for child development identified "care for the environment and nourishment of the senses" as going hand-in-hand. Having the right physical environment which provides sensory experiences, can "deeply affect the children's physical and emotional development," and have lifelong effects (Nicol and Taplin, 2012, p. 15). Across literature, sensory experiences are also encouraged through use of the outdoor environment.

Our study school is one of the few school sites in the UAE with green space, although they do have the challenge of searing heat, which often makes the use of that space untenable. It is only during the cooler months that doors can be left open and free flow encouraged. Large construction materials are found outdoors, all donated from the local community, and the learning opportunities afforded by these objects was enthusiastically described by the practitioners. For example, various size logs were recognised as supporting children with everything from sorting, preparing children for discrimination between letters and numbers, to core strength. But, of course, the use of such materials, scavenged or found, has its origins elsewhere, in Loose Parts.

Nicholson's theory of Loose Parts, developed in the late 60s and early 70s, identifies 'variables' (random objects) as aids for creativity and investigation. Nicholson (1971) argued that all objects have the potential to be creative, and that the possibilities of that creativity are only limited by the range of objects available to create with. Nicholson, a trained architect, spurned a society that taught children that only those qualified, and with expertise could create, be that buildings, music or art. Creativity, he stressed, should be available to all. As should responsibility, decision-making and challenge. Like Montessori, Steiner, Reggio Emilia, Deleuze and Guattari and numerous others ECEC theorists before him, Nicholson rejected conformity and the acceptance of existing norms as stifling. A Loose Parts approach to play encourages children to not simply repeat what has gone before, but to imagine something new, something different.

As well as encouraging autonomous thinking, the practitioners in UAE recognised how the use of Loose Parts made from natural and readily found materials could encourage children to respect resources, understanding that all objects have a use, and that items should be handled with care. Warden (2022) expands on this by introducing 'nature pedagogy,' an approach that integrates sustainability into educational practices. This method emphasises ecocentric values, placing nature at the core of decision-making and encouraging a deeper connection with the natural world. This moves beyond issues of cost, or avoidance of waste, to recognising the long-term impacts that our decisions have upon our planet. And the responsibilities that we all hold as custodians of that planet.

Conclusion

If a setting found in a country like the UAE, one of the lowest ranked for sustainability, can make steps toward a more responsible future, then imagine the strides that settings in England could make. The CA has provided this school with many accessible routes to a more sustainable future. It has introduced small steps to support children in fostering care and love for their planet, with practitioners creating a sustainably resourced learning environment and modelling a culture of appreciation for the planet and, to some extent, each other. But relationships remain a relatively unexplored aspect within our data so far. Warden (2022, p. 9) explains why the centrality of emotional connection, of *being*, also plays a key role in our children's development:

> It [goes] beyond the ecological practices of litter management, power use and ethically sourced materials to a place where it was a way of being with children that accepted the observable loose materials, such as leaves, sticks and stones, and the unobservable phenomena that occurs when we feel emotionally connected to a place.

Warden (2022) highlights how nature pedagogy is about how we work with children and build relationships which are value driven, and that these relationships can be with places and objects, not just people. Doull and Ogier (2024, p. 34), likewise, encourage pedagogues to foster a love and care towards nature.

Undoubtedly the CA prompts practitioners to think about their pedagogical practices and about their attitude towards the learning environment. It invites practitioners to take small steps to support children in fostering care and love for the planet, through practitioners creating a sustainably resourced learning environment which breaks down barriers between the child and nature. But perhaps relationships, and connectedness are areas that deserve further exploration in nature-focused settings. Malaguzzi (Reggio Emilia Approach, 2022) refers to this as 'participation,' and stresses that this is vital for any real change:

> Participation is the educational strategy that is constructed and lived in encounter and relations day after day. Participation valorises and makes use of the hundred languages of children and human beings, understood as plurality of points of view and of cultures.
>
> Participation generates and informs the feelings and culture of solidarity, responsibility and inclusion, and produces change and new cultures.

Our research with this international school in the UAE provided valuable insights into the implementation of the CA and its integration of sustainability within early years education. Perhaps because of the UAE's extremely high living standards, it faces significant sustainability challenges, making this exploration particularly relevant. The CA's emphasis on fostering autonomous, environmentally conscious children, aligns with global efforts to address climate change, as highlighted during COP28. Our findings suggest that the CA, with its roots in various pedagogical traditions, offers a promising framework for developing sustainable learning environments. By revisiting and revitalising these historical educational values, we can better prepare future generations to be proactive, responsible stewards of our planet.

Acknowledgements

We would like to extend our heartfelt gratitude to the practitioners from our case study international school in the United Arab Emirates. Their generous contributions of time, insights, and expertise have been invaluable in enhancing our understanding of The Curiosity Approach in practice.

References

Baloch, R. M., Maesano, C. N., Christoffersen, J., Mandin, C., Csobod, E., Fernandes, E. O., Annesi-Maesano, I. and Consortium, O. B. O. T. S. (2020) Daylight and School Performance in European Schoolchildren. *International Journal of Environmental Research and Public Health*, *18*(1), 258. https://doi.org/10.3390/ijerph18010258.

Carnie, F. (2003) *Alternative Approaches to Education: A Guide for Parents and Teachers*. London: Routledge Falmer.

Climate Change Performance Index (CCPI) (2024) *United Arab Emirates*. Available at: https://ccpi.org/country/are/ (Accessed 4 April 2024).

Cooperrider, D. L. and Srivastva, S. (1987) Appreciative Inquiry in Organizational Life. In Woodman, R. W. and Pasmore, W. A. (Eds.), *Research in Organizational Change and Development*, 129–169. Stamford, CT: JAI Press.

Davis, J. and Elliott, S. (2014) *Research in Early Childhood Education for Sustainability: International Perspectives and Provocations*. London: Routledge.

Deleuze, G. and Guattari, F. (1987) reprinted 2013. *A Thousand Plateaus*. London: Bloomsbury.

Department for Education (DfE) (2024) *Early Years Foundation Stage Statutory Framework*. Available at: https://assets.publishing.service.gov.uk/media/670fa42a30536cb92748328f/EYFS_statutory_framework_for_group_and_school_-_based_providers.pdf (Accessed 5 November 2024).

Doull, K. and Ogier, S. (Eds.) (2024) *Teaching Climate Change and Sustainability in the Primary Curriculum*. London: Learning Matters.

Feez, S. (2010) *Montessori and Early Childhood: A Guide for Students*. London: Sage Publications.

Hellyn, L. and Bennett, S. (2019) *The Curiosity Approach Presents: From Ordinary to Extraordinary*. The Curiosity Approach.

Isaacs, B. (2018) *Understanding the Montessori Approach: Early Years Education in Practice*. 2nd edition. Oxon: Routledge.

Malaguzzi, L. (1990) At international conference '"Who Am I Then? Tell Me That First". Knowledges in Dialogue to Guarantee Citizenship to the Rights and Potentials of Children and Adults', *Reggio Emilia*, March 1990 (transcript from the original audio recording by the Centro Documentazione e Ricerca Educativa).

Maslow, A. H. (1943) A Theory of Human Motivation. *Psychological Review, 50*(4), 370–396.

Ministry of Education (2017) *Te Whāriki: Early Childhood Curriculum*. Available at: https://www.education.govt.nz/assets/Documents/Early-Childhood/ELS-Te-Whariki-Early-Childhood-Curriculum-ENG-Web.pdf (Accessed 25 March 2025). New Zealand Curriculum.

Montessori, M. (1964) *The Montessori Method: Scientific Pedagogy as Applied to Child Education in "The Children's Houses"*. Massachusetts: Robert Bentley Inc.

Nicholson, S. (1971) How NOT to Cheat Children: The Theory of Loose Parts. *Landscape Architecture*, 62(1), 30–34. JSTOR, http://www.jstor.org/stable/44663886. Accessed 25 Mar. 2025.

Nicol, J. and Taplin, J. T. (2012) *Understanding the Steiner Waldorf Approach: Early Years Education in Practice*. Oxon: Routledge.

Office for National Statistics (ONS) (2022) *How Our Spending Has Changed since the End of Coronavirus (COVID-19) Restrictions*. Available at: https://www.ons.gov.uk/businessindustryandtrade/retailindustry/articles/howourspendinghaschangedsincetheendofcoronaviruscovid19restrictions/2022-07-11 (Accessed 4 April 2024).

Reggio Emilia Approach (2022) *Reggio Emilia Approach*. Available at: https://www.reggiochildren.it/en/reggio-emilia-approach/ (Accessed 4 April 2024).

Schaefer, C. E., and Kaduson, H. G. (Eds.) (2006) *Contemporary Play Therapy: Theory, Research, and Practice*. New York: Guilford Press.

Shaw, N., Eschenbrenner, B. and Baier, D. (2022) Online Shopping Continuance after COVID-19: A Comparison of Canada, Germany and the United States. *Journal of Retailing and Consumer Services*, 69, 103100. https://doi.org/10.1016/j.jretconser.2022.103100.

The Curiosity Approach (2024a) *About Us: Meet the Directors*. Available at: https://www.thecuriosityapproach.com/ (Accessed 4 April 2024).

The Curiosity Approach (2024b) *Sustainability in Early Childhood Education*. Available at: https://www.thecuriosityapproach.com/blog/Sustainability (Accessed 4 April 2024).

Thornton, L. and Brunton, P. (2014) *Bringing the Reggio Approach to Your Early Years Practice*. 3rd edition. Oxon: Routledge.

Thornton, L. and Brunton, P. (2015) *Understanding the Reggio Approach: Early Years Education in Practice*. 3rd edition. London: Routledge.

Warden, C. (2022) *Green Teaching: Nature Pedagogies for Climate Change and Sustainability*. London: Sage Publications.

Wilkinson, R. (1993) *Rudolf Steiner on Education: A Compendium*. Gloucestershire: Hawthorn Press.

Williams-Siegfredsen, J. (2017) *Understanding the Danish Forest School Approach: Early Years Education in Practice*. 2nd edition. Oxon: Routledge.

7

SUSTAINABILITY AND FOOD

A social justice perspective

Nicola Watson and Janet Harvell

Introduction

Fossil fuels are well-known as drivers of climate change and biodiversity loss but food production and consumption are also key drivers which have until now been largely overlooked (Morren et al., 2021). Awareness about what we eat and its impact is now receiving greater attention and in 2023, the United Nations Climate Change Conference (CoP 28) held, for the first time, a day dedicated to the impact of farming on climate change. In this chapter, we have chosen to focus on the UK, China and Tanzania, in order to reflect the commonalities that affect all peoples, communities and nations, but also the differing impacts as a result of geography, wealth and poverty. We recognise the discourse of blame which divides us; that is the suggestion that the actions of individuals, communities and smaller nations are irrelevant, while larger nations such as China, India and the USA are disproportionately responsible for climate change. We reject this view, not least because as educators, it would be unethical to convey to children and young people (CaYP), that their actions are meaningless and their influence is negligible. Although countries are not affected in the same way, our predicament is global rather than confined to individual nations. This recognition helps us to confront and challenge the notion that wealthier nations such as the UK (which is arguably, responsible for proportionally lower greenhouse gas emissions), have less responsibility to reduce them further. We argue that the national conversation on climate change should highlight the need to learn with and from other nations while critically evaluating our own cultural practices and aligning our actions with our shared values.

DOI: 10.4324/9781003506850-10

Although our relationship with food is largely culturally ingrained, there is evidence which suggests that when people have access to reliable information and engage in dialogue within their communities, they are empowered to make positive change. A good example of this is the People's Plan for Nature (2023) report produced by a UK citizens' assembly comprising a diverse range of ages, heritage, social classes and education. They were given a collective voice in order to make recommendations to government on protecting and renewing nature. Their final report included recommendations on food production and consumption and called for "the majority of people to change their diet for the better and aim to reduce meat, dairy and fish consumption by at least 25% by 2030" (PPFN, 2023, p. 33). This suggests that when people are informed, they are empowered to make positive change and to advocate for a sustainability policy agenda at national level. It is in this spirit that as educators we can be instrumental in the empowerment of CaYP to make informed decisions and have their voices heard at all levels of society.

We know that human actions are the cause of our predicament and this chapter considers the environmental impact of food production, consumption and the resulting exacerbation of inequalities on the poor both globally and within rich societies. Our comparison of China and the UK aims to illustrate their similar issues in terms of child health but also serves to highlight the effects of climate change on China in the present day. A case study on Tanzania is used to illustrate the impact upon a sub-Saharan African nation, one of the areas most affected by the climate crisis. We untangle some of the complexities, dilemmas and confusion facing educators today to address how we might support CaYP's empowerment to eat more sustainably and develop a healthy relationship with food. A case study from England exemplifies how a particular setting brought together parents and professionals to achieve this aim.

The environmental impact of food

Our globalised food system currently produces more than enough calories to feed the world, but not to provide a nutritionally balanced diet for everyone (KC et al., 2018, p. 6). According to the United Nations Environment Programme (2020) "Worldwide, more than 10 per cent of people are hungry, roughly 25 per cent are overweight or obese, and another 25 per cent – who may also be either underfed or overfed – are micronutrient deficient." While there have always been disparities in access to nutrition, caused by factors such as wealth and poverty, geography and societal structures, our food production and consumption patterns represent a vicious circle by driving climate change and biodiversity loss thereby worsening inequalities. According to the Intergovernmental Panel on Climate Change (IPCC, 2019) the food system is responsible for between 21% and 37% of carbon emissions. Of these, beef is a major culprit (Poore and Nemecek, 2018). Cattle produce methane which increases global warming at a far greater rate than carbon dioxide (Wood, 2023). There is currently a great deal of research aiming to reduce methane emissions, but as McGregor et al. (2021, p. 1180) point out, though this aim aligns with the aim of achieving climate goals, there is an inherent contradiction in that "it aligns with the key driver of virtually all research on animal agriculture: the ongoing expansion of animal industries." They assert that efforts to reduce and mitigate climate change must be based on fairness and social justice. Currently, there is a grossly disproportionate number of farmed animals on Earth. Indeed, Ritchie and Roser (2024) confirm the startling statistics that of all the mammals on Earth, 34% are human, 62% are farmed animals and just 4% are wild. It is not difficult to imagine the positive impacts on land and water use, and reduction in emissions if 62% of mouths to feed were taken out of the equation. McGregor et al. (2021, p. 1181) highlight the need for a radical shift in thinking:

> Despite amassing an exhaustive array of information about the devastating effects that climate change is having on food security, land degradation and desertification, the IPCC, EAT-Lancet Commission and Greenpeace all advocate for balanced diets including smaller portions of meat, rather than plant-based diets. It is likely these recommendations translate into minimal changes for richer countries.... If healthy diet strategies are already failing to significantly influence behaviours to improve the health of individuals and populations, as evident in increasing rates of food-related illness, it is hard to see how a reinforcement of such strategies will significantly limit animal agriculture emissions. It leaves the animal-intense food systems of wealthier countries and communities barely challenged.

The emissions from food of wealthier countries are much greater. The US diet, for example causes more than twice the emissions of the global average diet, whereas the lowest-emission diets can emit ten times less (Bridle, 2020). It is important to recognise that environmental impact comprises more than carbon emissions. Morren et al. (2021, p. 1) explain that "Environmental impact is a diverse concept and is often quantified by using three indicators: carbon footprint (CO2 equivalent emissions), ecological footprint (land needed to regenerate resources) and water footprint (water needed to regenerate resources)." The increased production of farmed animals for food has had a very significant impact on all three. A stark illustration of the ecological impact our food production and consumption are having on the earth is provided by Dasgupta (2021, p. 1):

> Today, we ourselves, together with the livestock we rear for food, constitute 96% of the mass of all mammals on the planet...We are destroying biodiversity, the very characteristic that until recently enabled the natural world to flourish so abundantly. If we continue this damage, whole ecosystems will collapse.

Brewood and Garnett (2023, p. 24) concur on the benefits of reducing animal food consumption "in countries (primarily richer ones) where people already eat a lot of animal products." Aside from lower emissions and less use of water, "diets lower in animal products than today's diets would notably decrease grazing land use and also decrease arable land use, freeing up land that could potentially be used to restore ecosystems." In addition to the environmental impact of food production and consumption, waste is also a major issue. According to Dasgupta (2021, p. 36), "1.3 billion tonnes of food is lost or wasted every year," representing one third of all food produced.

The direct impact of climate change on weather, such as increased storms, heatwaves and flooding have been found to have a direct link to childhood mortality and ill health (Helldén et al., 2021). Save the Children (2023) reports that children are disproportionately affected by climate, particularly following drought, there is more emphasis on all members to earn additional income which risks adverse effects on health, education and wellbeing. For example, when households are experiencing food insecurity for the household. This could then reflect on children's ability to continue their education if families are starving and need all family members to engage in work activities (Bohn, 2022).

The last 70 years has seen China's average temperature increase by more than the global average, while a 2021 climate assessment acknowledged that coastal water levels were at their highest since 1980 as a result of glacial retreat and a decline in sea ice (Stanway, 2022). These changing weather patterns will obviously impact on the water availability and affect ecosystems and food production, with food insecurity expected to increase, with the inevitable impact on children's nutrition, wellbeing, and academic achievement (Helldén et al., 2021). Save the Children identifies that nearly every child alive can now expect to experience at least one climate-linked incident annually, with over 1 billion children in 2023 being at extreme risk because of climate-related incidents. This number is expected to increase to 4.2 billion by 2053 (UNICEF, 2021).

It is salutary to recognise that those countries contributing least to the problem, are those most at risk. Hayhoe (2022, p. 52) states that it is having most impact on those who have "done the least to contribute to the problem." This is illustrated in the case study below.

Case study – Tanzania: threats and opportunities

Baptista et al (2022) state that 123 million people in sub-Saharan Africa (SSA) (12% of the population) were predicted to experience acute food insecurity. Vulnerable countries are already struggling to attain the UN Sustainable Development Goal 2 (SDG2) related to food security, because of weather-related incidents linked to climate change i.e. droughts, floods, cyclones, with a third of droughts occurring in SSA.

Tanzania is an example of a country which has a low capacity for climate resilience, and a poor water security situation which is only expected to intensify. The Children Climate Risk Index (CCRI) identifies where children are most at risk, with Tanzania in the top 25% of countries which are already experiencing severe weather change because of climate change (UNICEF, 2021), and is identified as one of the 163 countries that are at "extremely high risk" that together contribute less than 10% of global transmissions. Indeed, between 2015 and 2016 drought and flooding had devastating effects on both livestock death and crop failures, with the lack of rainfall being acutely associated with food insecurity (Ires, 2021). This is unlikely to be a rare event with predictions suggesting that although temperatures will be warmer overall, it is the rainfall that will be particularly unpredictable with more rainfall in the north of the country, but less in the south (Bohn, 2022). This is particularly concerning as it has been identified that in Tanzania "...food insecurity increases by 5 to 20 percentage points with each drought or flood" (Baptista et al., 2022). Currently, 59% of the Tanzanian population cannot afford a nutritionally adequate diet and there is already evidence that climate change is having an impact on vulnerable families and children with the threat of increased malnutrition (Africa Health, 2016). This is worrying in a country which is predicted to become one of the most populous nations in Africa over the coming decades (British Council, 2016). Currently, 50% of the population are children, 32% of whom are under five, and 32% recognised as having stunted growth (an increase of 16% in five years), while 14% are underweight (UNICEF, 2023; US Aid, 2021). The role of education cannot be underestimated, with these rates decreasing as maternal education increases, nevertheless a recent UNICEF (2023) report identified that there is insufficient learning around climate crisis in schools, with children eager to learn more and understand how they can make a difference. At the same time, poor nutrition and economic disruption are likely to lower children's educational achievement or keep them out of school altogether.

There are initiatives aiming to address this urgent situation. For example, a five-year project, Building Climate Resilience in the Landscapes of the Kigoma Region of Tanzania, was recently approved by the Global Climate Fund in Tblisi (UNHCR, 2023, online). This is intended to be a flagship method focusing on using nature-based solutions including reinstatement of ecosystems, and is in an area that is also home to a large number of displaced people. It is intended that up to 570,000 people will benefit positively from the project, which will focus on supporting land use to be more climate resilient, while maintaining "261,000 hectares of forest and agroecological ecosystems in support of local livelihoods in the districts of Kasulu, Kakonko and Kibondo" (UNHCR, 2023, online).

Changing the narrative

Food education is often focused on nutrition, which is of course, vital to health. However, few educators are dieticians, and our knowledge of food tends to be drawn from varying dominant cultural discourses and is more complex than the meeting of nutritional requirements. Wilson (2015, p. 1) says how we eat is learned. As educators, we can influence the CaYP with whom we work. This involves supporting their development of knowledge and skills to enable them to make more informed choices. We recognise that this is challenging and requires the support of policy makers, communities and families. We have selected some key (though by no means exhaustive) issues, as a focus for bringing about positive change.

Critiquing the notion of choice

It is well known that we all need to eat a diet rich in fruit and vegetables and low in sugar and all national guidelines from around the world reflect this (KC et al., 2018). China is an example of a country that has undergone significant changes in recent history and is now classed as an upper-middle-income country. The impact of this on people's disposable income has resulted in enhanced choices, particularly the consumption of fast foods and animal-based products, but this comes at the expense of increased greenhouse gas emissions (Myers, 2022; Sheng and Jia, 2023). The impact of climate change on critical food supplies, together with improved living standards and a greater reliance on a more animal-based, highly processed diet, brings about particular challenges (Clark, 2022). Around 8.3% of children under 5 in China were overweight or obese in 2020 (Ye, 2023). This is similar to the UK where in 2022/2023 9.2% of four-to-five-year-old children were classed as obese, (RCPH, 2023). In China, the transition from predominantly plant-based, unprocessed foods, to diets containing highly processed foods rich in fats, sugar, and animal products has resulted in an increase in health problems such as diabetes and obesity (Myers, 2022). Conversely, increasing poverty in the UK can make access to nutritious, sustainable food limited or unavailable, removing or severely limiting choice. In her review of qualitative research on food and nutrition inequalities in the UK, Douglas (2023, p. 3) found that people lacking food security in the UK wanted to provide healthy food for their families but were constrained by the cost of food "in terms of the types and quantity of food" but also "by the necessary energy and other essential costs needed to procure, prepare and consume food safely, e.g. transportation, kitchen equipment and means of maintaining food hygiene in the home, e.g. hot water and washing facilities."

Morren et al. (2021, p. 1) notes the confusion resulting from the "multifaceted nature of sustainable food choices, and contradictory information on the footprint, [which] makes it hard for consumers to learn about and act upon the information available." It is worth noting a few examples to illustrate this here. For instance, it is easy to assume that the traditional "meat and two veg" familiar to the UK diet is sustainable and 'natural.' Bridle (2020, p. 198) points out that "worldwide, the amount of animal-based protein eaten per person has risen by over 50% since the 1970s" and that even ignoring land and water considerations, due to "greenhouse-gas emissions alone, we simply cannot scale up agriculture to satisfy consumer demand." Another example is the oft-seen suggestion that eating locally-produced food is a more important consideration than it is. Ritchie et al. (2020, online) explains that "Transport is a small contributor to emissions. For most food products, it accounts for less than 10%, and it's much smaller for the largest GHG [greenhouse gas] emitters. Similarly, livestock who are grass-fed are often regarded as 'environmentally-friendly.' However, "grass-fed beef and lamb has a huge carbon and land footprint compared to pork, chicken and protein rich plants such as peas, beans and lentils" (Garnett, 2023, online).

The vast majority of parents and educators want CaYP to eat healthily and may try to limit or proscribe foods high in fat, salt and sugar, but it is very difficult to avoid exposure to aggressively marketed foods manufactured to be hyper-palatable to CaYP (Van Tulleken, 2023). To complicate matters further, Bahadur, et al. (2018, p. 2) allege that even nutritional guidelines "may be vulnerable to political and industry interference."

Food and pedagogy

Major changes in the way that food is produced and consumed are necessary to ensure sustainability (Poore and Nemecek, 2018). Educators play an important role in informing and educating CaYP about their food practices and in helping them to develop healthy and sustainable relationships with food. Food impacts on every area of our lives and requires us to consider culture, economics, availability, nutritional requirements and psychological factors. Like obesity, poverty is often framed as a moral failing, and as Douglas (2023, p. 2) points out, "people who are living with poverty and obesity face a 'double whammy' in terms of societal disapproval and judgment about their status." Monterrosa et al. (2020, p. 60) point out that food practices are transmitted socially, primarily from parents to children and "are connected to social and economic expression of identities, preferences, and cultural meanings." It is crucial therefore, that educators are highly sensitive to social contexts and avoid judgemental or shaming discourses around food. Positive approaches to change are likely to be more effective in the long term than enforced directives or proscriptions. Access to reliable information within a supportive environment which foreground sustainability is key. Thaler and Sunstein (2008, p.6) developed the theory of 'nudge' as a strategy which they define as, "any choice architecture which alters people's behaviour in a predictable way." They stress that nudges are not mandates. "Putting the fruit at eye level counts as a nudge. Banning junk food does not." Morren et al. (2021, p. 11) researched the effects of 'nudges' to promote sustainable eating and their findings suggest that "even a single health information nudge can motivate people without dietary restrictions (i.e. meat-eaters) to change their diet."

Nudging is an example of using positive psychology to effect change. As the name suggests, positive psychology is an approach which focuses on strengths rather than weaknesses (Ackerman et al., 2018). Such a focus can help educators reframe dietary change away from depriving oneself of what is familiar towards learning with curiosity and exploration of food and eating. The NASA scientist and climate activist, Peter Kalmus displays reverence and appreciation of our planet's ability to feed us when he says:

> When I pick a leaf or a fruit from a plant, or even when I pull out a weed, I do so with awareness, respect, and gratitude. Every seed is precious, a tiny genetic blueprint of unimaginable complexity, evolved over billions of years, for a plant being that creates a miraculous gift: nourishing food that we can eat.

This reflection offers an expanded conception of food and multiple perspectives which can inspire learning opportunities for learning with head, heart and hands and an appreciation of the natural world as explored in other chapters of this book, particularly, Chapter 4. Obvious examples include projects around growing and cooking food where resources allow.

Clearly, initiatives depend upon cultural and sub-cultural contexts and will differ according to demographics. As has already been mentioned, identity plays a significant role in shaping food practices. An example of this is the association of meat-eating with masculinity. Mertens et al. (2023, p. 7) explain that a reduction in meat consumption may represent "a violation of the spirit of masculinity given its well-defined characteristics such as dominance orientation and being powerful and tough." However, discourses change and are changed by cultural influences. An example of this can be seen in the UK's men's football team, Forest Green Rovers (FGR). Football is a very popular and culturally influential sport around the world. It is strongly associated with masculinity, but "also offers a unique space for the public performance of identity" (Power et al., 2020, p. 1). FGR is owned by the climate activist Dale Vince and is the world's first carbon neutral club. It offers exclusively plant-based food at its grounds and has been described by the United Nations as the greenest football team in the world (United Nations Climate Change, 2020).

Working with communities offers the opportunity for the sharing of resources and the development of dialogue to achieve shared understandings.

Working *with* others in partnership rather than as experts seeks to ensure that individual and cultural identities are respected and included. The following case study represents an exemplar of how one setting worked with parents and young children with the aim of promoting a sustainable food offer.

Case study – Supporting one setting's transition to a more plant-based menu

Ailsa McHardy is a Paediatric Dietitian with a keen interest in sustainability. She suggests that Paediatric Dietitians should play an integral role in bespoke menu planning for settings which should reflect social and ethnic diversity. In 2022, Ailsa worked with an environmentally conscious setting in England who were aiming to transition to a more plant-based menu. The setting already had sustainable practices in place such as growing their own vegetables and learning about recycling. It was, therefore, a natural progression for them to look at the environmental impact of the setting's catering. Their aims were to provide nutritious, appealing food and reduce waste while working in partnership with parents and carers.

Ailsa highlighted the key nutrients in the existing menus and made suggestions for how these nutrients could be increased, as well as giving ideas for appropriate plant-based alternatives to some of the meat and fish dishes still present. Menus were adapted to include a list of 'Plant Power Points' – indicating the number of different plant foods being offered on that specific day – and a list of 'Today's Protein Stars,' to indicate the key sources of plant-based protein being provided each day. The idea with this approach was that it would both appeal to the children and be informative for the nursery team and their parents.

Simple menu swaps were discussed and Ailsa made suggestions for appropriately fortified plant-based milks with a more rounded nutrition profile than the choices already in place and helped identify which plant-based meat alternatives, such as sausages, were most 'child-friendly' in terms of their protein, fat and salt contents.

Central to the project was working with parents in making the changes. Ailsa explained, "We discussed some of the questions which parents had already voiced about the plans for a plant-based menu and, as expected, these included concerns around where the children would get protein and iron from in the meals, and what would happen if a child didn't like the plant-based options. We opted to produce a 'parent information' sheet, which included information on *why* a more plant-based approach was being adopted and also addressed some frequently asked questions, for example "is soya safe for children?" and "where will they get protein from?." After a few minor tweaks based on feedback from the nursery team, the parent information sheet was finalised and references to the evidence-base around plant-based diets included."

The feedback from the setting showed that the evolving plant-based menus were popular and the nursery team commented that the parents of children just starting at the nursery especially liked how much plant-based produce was on offer.

Adapted with permission from Ailsa McHardy (2022)

Conclusion

Animal agriculture is accelerating the destruction of the planet, our only home. Individual actions alone, although essential, are not enough. Policy makers need to enact the findings from research on the drivers of climate change and loss of biodiversity but as McGregor et al. (2021, p. 1182) point out,

> Research communities who are genuinely concerned about reducing emissions from animal agriculture and produce the knowledge that inform the biopolitical strategies promoted by global bodies like the IPCC [International Panel on Climate Change] and the FAO [Food and Agriculture Organisation] need to prioritise justice, rather than efficiency, as the core principle to steer a radical transition towards low emissions food systems.

As educators we can play a role in instilling and shaping attitudes to food and thereby educate for sustainability. There is evidence to show that in the UK, meat consumption is declining (Stewart et al., 2021) which has positive implications for health as well as climate change and biodiversity loss, although this needs to accelerate to meet the UK's target for net zero by 2050 (Climate Change Committee 2020). As individuals, groups and communities, exercising choices in ways which align with individual and collective ethical values can be empowering. Rather than denying oneself familiar foods, learning more about health, nutrition, animal welfare and broadening our culinary repertoire can be a deeply rewarding form of activism in the global struggle for social justice for current and future generations. In the fight against climate change and loss of biodiversity, our relationship with food needs a more visible place at the table.

References

Ackerman, C.E., Warren, M.A. and Donaldson, S.I. (2018) 'Scaling the heights of positive psychology: A systematic review of measurement scales', *International Journal of Wellbeing*, 8(2), pp. 1–21. Available at: https://doi.org/10.5502/ijw.v8i2.734.

Africa Health (2016) Climate Change Forces Tanzanian Households to Cut Meals. Available at: https://africa-health.com/research/climate-change-forces-tanzanian-households-cut-meals/ (Accessed 24 January 2024).

Baptista, D.M.,Farid, M., Fayad, D., Kemoe,L., et al. (2022) Climate Change and Chronic Food Insecurity in Sub-Saharan Africa. Available at: https://www.elibrary.imf.org/view/journals/087/2022/016/article-A001-en.xml (Accessed 24 January 2024).

Bohn, K. (2022) Lower Than Normal Rainfall Linked with a Higher Chance of Food Insecurity. Available at: https://www.preventionweb.net/news/lower-normal-rainfall-linked-higher-chance-food-insecurity#:~:text=Specifically%2C%20the%20researchers%20found%20that,less%20rain%20on%20maize%20production (Accessed 20 February 2024).

Bridle, S. L. (2020) *Food and Climate Change without the Hot Air.* Cambridge: UIT.

British Council (2016) Tanzania's Next Generation. Available at: https://www.britishcouncil.org/research-insight/tanzania-next-generation#:~:text=populations%20on%20Earth-,Tanzania%20is%20a%20country%20of%2055%20million%20people%2C%20of%20whom,it%20one%20of%20Africa's%20largest%20 (Accessed 20 February 2024).

Clark, M. (2022) 'The calorie question ', in *The climate book*. Edited by G. Thunberg. Allen Lane , pp. 131–133.

Climate Change Committee (2020) *The UK's Net Zero target The UK's independent adviser on tackling climate change* (no date). Available at: https://www.theccc.org.uk/wp-content/uploads/2020/10/CCC-Insights-Briefing-3-The-UKs-Net-Zero-target.pdf.

Dasgupta, P. (2021) *The Economics of Biodiversity: The Dasgupta Review*. Abridged Version. London: HM Treasury.

Douglas , F. (2023) 'What qualitative research can tell us about food and nutrition security in the UK and why we should pay attention to what it is telling us', in *The Nutrition Society. Conference on 'Diet and health inequalities'*, Cambridge University Press, pp. 1–10.Available at: https://doi.org/10.1017/s0029665123003713.

Garnett, E. (2023) Five Misused Food and Farming Terms, from Natural to Intensive – and What They Really Mean, *The Conversation.* Available at: https://theconversation.com/five-misused-food-and-farming-terms-from-natural-to-intensive-and-what-they-really-mean-200797 (Accessed 21 March 2025).

Helldén, D., Andersson, C., Nilsson, M., Ebi, K.L., Friberg, P. and Alfvén, T. (2021) 'Climate change and child health: A scoping review and an expanded conceptual framework', *Lancet Planet Health*, 5(3), pp. 164–175.

Hayhoe (2022, p. 52) Hayhoe, K (2022). Heat. In Thunberg, G (2022) *The Climate Book*. Allen Lane, UK.

IPCC (2020) An IPCC Special Report on Climate change, desertification, Land degradation, Sustainable Land management, Food security, and Greenhouse Gas Fluxes in Terrestrial Ecosystems Climate Change and Land Summary for Policymakers WG I WG II WG III. Available at: https://www.ipcc.ch/site/assets/uploads/sites/4/2020/02/SPM_Updated-Jan20.pdf.

Ires, I., 2021. Intensive agriculture as climate change adaptation? economic and environmental tradeoffs in securing rural livelihoods in Tanzanian River Basins. *Frontiers in Environmental Science*, 9, p.674363.

Kalmus, P. (2017) Being the Change: Live Well and Spark a Climate Revolution. Available at: https://peterkalmus.net/books/read-by-chapter-being-the-change/being-the-change-chapter-12-reconnecting-with-mother-earth/ (Accessed 31 October 2024).

KC, K.B. et al. (2018) 'When too much isn't enough: Does current food production meet global nutritional needs?', PLOS ONE. Edited by P.C. Struik, 13(10), p. e0205683. Available at: https://doi.org/10.1371/journal.pone.0205683.

Lu, C. (2023) Climate Change Could Drown China's Food Security. Available at: https://foreignpolicy.com/2023/09/08/china-climate-change-food-security-flood-agriculture/#:~:text=It's%20not%20just%20floodwaters%20that,wheat%2C%20corn%2C%20and%20soybeans (Accessed 9 January 2023).

McGregor, A., Rickards, L., Houston, D., Goodman, M., et al. (2021) 'The Biopolitics of Cattle Methane Emissions Reduction: Governing Life in a Time of Climate Change', *Antipode*, 53(4). pp. 1161–1185. Available at: https://doi.org/10.1111/anti.12714.

Mertens, A. and Oberhoff, L. (2023) 'Meat-eating justification when gender identity is threatened – The association between meat and male masculinity', *Food Quality and Preference*, 104, p. 104731. Available at: https://doi.org/10.1016/j.foodqual.2022.104731.

Monterrosa, E.C., Frongillo, E.A., Drewnowski, A., de Pee, S., et al. (2020) 'Sociocultural Influences on Food Choices and Implications for Sustainable Healthy Diets', *Food and Nutrition Bulletin*, 41(2), pp. 59S73S. Available at: https://doi.org/10.1177/0379572120975874.

Morren, M. Mol, J.M., Blasch, J.E., and Malek, Z. (2021) 'Changing diets – Testing the impact of knowledge and information nudges on sustainable dietary choices', *Journal of Environmental Psychology*, 75, p. 101610. Available at: https://doi.org/10.1016/j.jenvp.2021.101610.

Myers, S. (2022) Food and Nutrition. In Thunberg, G. (ed.) *The Climate Book*. Allen Lane, UK.

Nature Climate Change (2023) 'Modern food emissions', *Nature Climate Change*, 13(3), pp. 205–205. Available at: https://doi.org/10.1038/s41558-023-01643-2.

People's Plan for Nature (2023). Available at: https://peoplesplanfornature.org/ (Accessed 22 February 2025.

Poore, J. and Nemecek, T. (2018) 'Reducing food's environmental impacts through producers and consumers', *Science*, 360(6392), pp. 987–992. Available at: https://doi.org/10.1126/science.aaq0216.

Power, M.J. Widdop, P., Parnell, D., Carr, J. and Millar,R. (2020) 'Football and politics: the politics of football', *Managing Sport and Leisure*, 25(1–2), pp. 1–5. Available at: https://doi.org/10.1080/23750472.2020.1723437.

Ritchie, H. (2020) You Want to Reduce the Carbon Footprint of Your Food? Focus on What You Eat, Not Whether Your Food Is Local, Our World in Data. Available at: https://ourworldindata.org/food-choice-vs-eating-local.

Ritchie, H. and Roser, M. (2024) Wild Mammals Make Up Only a few Percent of the World's Mammals. *Our World in Data* [Preprint]. Available at: https://ourworldindata.org/wild-mammals-birds-biomass#article-citation.

Royal College of Paediatrics and Child Health (2023) *RCPCH responds to latest childhood obesity figures for England, 2022/23, RCPCH*. Available at: https://www.rcpch.ac.uk/news-events/news/rcpch-responds-latest-childhood-obesity-figures-england-202223.

Save the children (2023) Tackling the Climate and Inequality Crisis for Children Tanzania Data Snapshot. Available at: https://resourcecentre.savethechildren.net/pdf/StC_spotlight_Tanzania_AW-2.pdf/ (Accessed 20 January 2024).

Save the Children (2024) How Is the Climate Change Impacting on Children? Available at: https://www.savethechildren.net/Climate-Crisis-Impact-Chidren (Accessed 20 February 2024).

Sheng, Y. and Jia, S. (2023) Climate Change Threatens China's Food Security. Available at: https://www.eastasiaforum.org/2023/09/02/climate-change-threatens-chinas-food-security/#:~:text=Recent%20studies%20show%20that%20extreme,droughts%20and%20rising%20carbon%20emissions (Accessed 9 January 2024).

Stanway, D. (2022) China Warns That Its Temperatures Are Rising Faster Than the Global Average. Available at: https://www.reuters.com/world/china/china-warns-that-its-temperatures-are-rising-faster-than-global-average-2022-08-04/ (Accessed 9 January 2024).

Stewart, C., Piernas, C., Cook, B. and Jebb, S. A. (2021) 'Trends in UK meat consumption: analysis of data from years 1–11 (2008–09 to 2018–19) of the National Diet and Nutrition Survey rolling programme', *The Lancet Planetary Health*, 5(10), pp. e699–e708. Available at: https://doi.org/10.1016/S2542-5196(21)00228-X.

Thaler, R. and Sunstein, C. (2008) *Nudge: Improving Decisions about Health, Wealth, and Happiness*. New Haven: Yale University Press.

UNHCR (2023) Tanzania Builds Climate Resilience in Major Push to Restore Landscapes Hosting Displaced Populations. Available at: https://www.unhcr.org/africa/news/press-releases/tanzania-builds-climate-resilience-major-push-restore-landscapes-hosting#:~:text=Tbilisi%2C%2025%20October%202023%20%E2%80%93%20The,UNHCR%2C%20the%20UN%20Refugee%20Agency (Accessed 21 February 2024).

UNICEF (2021) The Climate Crisis is a Child's Rights Crisis. Available at: https://www.unicef.org/media/105531/file/UNICEF_climate%20crisis_child_rights_crisis-summary.pdf (Accessed 24 January 2024).

UNICEF (2023) Children for Climate Action: Voices from Tanzania. Available at: https://www.unicef.org/tanzania/media/3471/file/Children%20for%20Climate%20Action%20Summary%20Report.pdf (Accessed 21 February 2024).

United Nations Climate Change (2020) Creating the Greenest Football Club in the World – Forest Green Rovers | United Kingdom, Unfccc.int. Available at: https://unfccc.int/climate-action/momentum-for-change/climate-neutral-now/creating-the-greenest-football-club-in-the-world-forest-green-rovers.

United Nations Environment Programme (2020) How to Feed 10 Billion People, UNEP. Available at: https://www.unep.org/news-and-stories/story/how-feed-10-billion-people.

US Aid (2021) Tanzania Nutrition Profile. Available at: chrome-extension://efaidnbmnnnibpcajpcglclefindmkaj/https://www.advancingnutrition.org/sites/default/files/2023-05/usaid_an_tanzania_final_2023.pdf (Accessed 24 January 2024).

Van Tulleken, C. (2023) *Ultra-Processed People*. London: Random House.

Wilson, B. (2015) *First Bite: How We Learn to Eat*. London: Fourth Estate.

Wood, J. (2023) Curbing Methane Emissions Is Key to Slowing Climate Change, World Economic Forum. Available at: https://www.weforum.org/stories/2023/10/tackling-methane-levels-is-the-quickest-way-to-slow-climate-change-say-scientists/ (Accessed 31 October 2024).

Zhang, J.F., Mauzerall, D.L., Zhu, T., Liang, S., Ezzati, M. and Remais, J.V. (2010) 'Environmental health in China: Progress towards clean air and safe water', *Lancet*, 375(9720), pp. 1110–1119. [PMC free article].

Zhu, Z. and Yin, P. (2023) 'Overwieght and obesity: The serious challenge faced by Chinese children and adolescents', *Journal of Global Health*, 13, p. 03036. Available at: https://www.ncbi.nlm.nih.gov/pmc/articles/PMC10357130/#R3 (Accessed 9 January 2024).

8

NATURE CONNECTEDNESS AND THE COMMON THIRD WITHIN THE PRIMARY CURRICULUM – EXPLORING SUSTAINABILITY THROUGH A CROSS-CURRICULAR APPROACH

Rosemarie Hill and Dan Hughes

Introduction

During the night of 27th September 2023, a brutal attack took place in England that shook the public consciousness. The destruction of a solitary tree sent shockwaves through the country and beyond. People came together to mourn the loss of a tree just as they would mourn a friend who had recently passed away. Every day, trees are destroyed to make way for different aspects of human life; housing, farming, industry and entertainment. Yet, this one tree brought everyone together; a tree that connected people with each other and with nature.

The Sycamore Gap tree, standing proudly in a dip along Hadrian's Wall for over 100 years, was felled overnight to widespread condemnation (Robinson, 2023). This was a senseless act with no apparent motive. The public response was overwhelmingly united in seeking justice and wanting to know why. It required us to consider our connection with this tree and nature more widely – why was this so upsetting? Why did it appear on worldwide news? What has this taught us about nature and our relationship with it? How can we maintain, develop and strengthen this connection?

Nature connectedness lies at the heart of being human (Pritchard *et al.*, 2019); this powerful event illustrates our deep-set, core beliefs around the value of life and nature – not just human life, but life in all its forms. As educators, we believe that nature should be at the forefront of our thinking and that education has a role to play in strengthening our connection to nature and each other. Nature and people connectedness has the power to change how young people view the world and their place in it. The social pedagogical concept of the Common Third, which will be explored later in

DOI: 10.4324/9781003506850-11

this chapter, is a potential vehicle for this transformational change. In this chapter we will explore ways in which schools can help children to become more connected to their wider world.

Nature (dis)connectedness

Being connected to nature has been at the core of human existence since the beginning of time (Pritchard *et al.*, 2019). As humans we are drawn to nature, from the people living in the Stone Age using nature for survival through to the more modern movements highlighting and appreciating the beauty of nature. Humanity's relationship with nature has been explored by writers, academics and explorers throughout history. William Morris (1834–1896) could be seen as one example. A textile designer, artist, poet and social activist, Morris became an influential contributor to the Arts and Crafts movement, where nature was used as inspiration for textile and furniture designs as well as works of art. However, this is not only about aesthetics, this relationship, connection or friendship with nature is necessary for our very survival, as well as for supporting our mental health and well-being. We have a need to connect with our natural world, an inherent need to connect with life in all its forms (Jones, 2020).

Nan Shepherd (1893–1981) a twentieth century writer, poet and mountaineer, defines this relationship in her influential mediation on nature, The Living Mountain. Instead of the traditional attitude of conquering and exploiting nature, Nan describes a friendship with nature. She walks to "merely to be with the mountain as one visits a friend with no intention but to be with him" (Shepherd, 2011, p. 15). It is this concept of friendship with nature and how it could be the foundations of nature connectedness that we will seek to explore further in this chapter.

Nan Shepherd is one of many who was fascinated with and attempted to tie down humans' relationships with nature. Edward Wilson, in his fascination with human's relationship with nature, coined the term biophilia (Wilson, 1984). Biophilia stems from the Latin: bio meaning life and philia meaning love. In his 1984 text Wilson focused on the joy we humans feel when we see or hear something beautiful or wonderous in nature. When most people see a murmuration of starlings or listen to the rustling of wind-blown leaves, they experience biophilia. Nan Shepherd was experiencing biophilia when she walked in the Cairngorms with the only intention of being there. However, she called it elemental – being at one and immersed with the living mountain all round her. This is nature connectedness: being at one with nature, being immersed in nature, noticing and appreciating the awe and wonder nature creates in us.

There is compelling evidence from scientific research which suggests that increased exposure to green spaces can have a positive impact on blood pressure, hypertension, asthma and coronary heart disease in adults (Twohig-Bennett and Jones, 2018). The Japanese practice of forest bathing (Shinrinyoku, date), which involves breathing in the phytoncides (wood essential oils) produced by trees, can increase human immune function (Quing Li, 2010). Quing Li's research also highlights the restorative benefits of being in the 'the quiet atmosphere, beautiful scenery, mild climate, and clean fresh air in forests' (2010, p. 9). The implication is the benefits are not only physiological but also psychological, improving and calming anxiety. What is more, research shows that only one forest bathing trip a month is necessary to feel these benefits. In the UK, science validates the ancient Celtic prescription of walking in forests to support breathing and to eliminate the lungs of toxins caused by colds and influenza (Beresford-Kroeger, 2019, p. 183). Another study has found that a bacteria found in soil (Mycobacterium vaccae) could increase serotonin levels in the brain as well as improving the quality of life for cancer patients (Lowry *et al.*, 2007). Further studies into this bacterium suggest it could improve conditions such as asthma and colitis as well as improving mental health (Smith *et al.*, 2019). It turns out that making mudpies could be good for our health. Finally, nature is even thought to impact upon our very character; studies indicate that immersion in nature can make us more generous and caring (Weinstein *et al.*, 2009). Nature impacts upon us as humans, on multiple levels.

Applying the lens of nature connectedness to children in particular, Wells' (2000) longitudinal study suggests that children's cognitive function increased after they were relocated to a home with more green spaces nearby. In 2015, a study found that children with a higher connection to nature had higher attainment in English (Richardson *et al.*, 2015). Ingvar (cited in Jones, 2020, p. 56) explains the impact that nature can have on the developing child, sharing: "our brains [need] to be stimulated from the flow of sound, light, shapes and colours that nature provides, especially between the ages of three to six, when the energy flow in the human brain is at its greatest."

We need nature, from both a physical and psychological perspective as well as from a cultural and survival perspective. Humans need a connection to nature to survive, to learn and to thrive. However, our relationship with nature has become increasingly exploitative. We are at a time of climate crisis due to the destructive actions of humans upon the planet. The average world temperature is rising: 2023 was the warmest year since records began (World Meteorological Organisation, 2024). The 2023 State of Nature Report found a significant loss in the UK's plants, animal and fungi since 1970 (State of Nature Partnership, 2023). At the same time, we are even losing the words to talk about the natural world. In 2007, there was a purge on nature words in the Oxford Junior Dictionary such as almond, blackberry and crocus in favour of words such as analogue, block graph and celebrity (Flood, 2015). Further studies indicate that three-quarters of children spend less time outdoors than prison inmates (Carrington, 2016). To measure this apparent disconnection from nature, the Nature Connectedness Index (NCI) was developed. This is a quantitative measure of both children's and adults' connection to nature over their lifespan and developed from analysis of UK National Statistics, in particular the Monitor of Engagement with the Natural Environment Survey (MENE) (Richardson *et al.*, 2019). Alarmingly, these statistics shows how nature connectedness begins to fall from the ten- to twelve-year-old bracket and only regains previous levels when participants reached the 31–40-year-old bracket. (Richardson *et al.*, 2019, p. 7). Armstrong (2022) reasons that this disconnection is hardly surprising, as when children do venture into the outdoors, they will often view the world through a digital lens, more concerned with the perfect photo for social media than being present and at one with nature. As educators it is important to maintain hope, and there are actions that we can take to strengthen children's connection with nature and in turn support "nature's revival" (Richardson *et al.*, 2015, p. 7).

Reconnecting

The outrage caused by the Sycamore Gap tree incident suggests our emotional connection with nature endures, despite being hidden under many layers of urbanisation and technology. Among children, this concern for the natural environment is more apparent, with eight in ten children agreeing that looking after the environment was important to them (Natural England, 2021). There is hope. We can rekindle our friendship with nature, starting with our children. Jones (2020) contends that if a child is introduced to nature before the age of 12, there is an increased chance they will continue that relationship and connection into adulthood. Therefore, by placing nature "at the heart of children's school education" we can develop and deepen our children's relationship with nature (Richardson *et al.*, 2015, p. 21).

The Sustainability and Climate Change strategy (SCCS, DfE, 2023) was designed as a starting point for educators to begin to rekindle children's connection to nature and to support the restoration of the environment through education. The strategy aims to support the United Nations Sustainable Development Goals (SDGs) and advises the teaching of sustainability through a knowledge-rich approach, while stating that children "must also be given the hope that they can be agents of change, through hands-on activity" (DfE, 2023, n.p). On the surface this appears to be a step forward in promoting nature connectedness within educational settings. However, the draft strategy received criticism, with Rushton and Dunlop (2022, n.p.) referring to it as a "placebo policy" doing no more than "repurposing and reframing of exiting policy." The emphasis of the SCCS does appear to rest on knowledge and learning about the causes of climate change and its solutions, however, there appears to be only a little detail regarding empowering children to develop a physical, psychological, spiritual and emotional connection with nature. Although words such as "holistic" and "practical activities" are mentioned, these appear as tokenistic within a fundamentally performative framework.

While initiatives such as the National Education Nature Park and the Climate Leaders Award are discussed within the SCSS, with an acknowledgement that not all children and young people have equal access to green spaces, the details of how such initiatives will be funded and implemented are somewhat hazy. Barriers such as limited funding, lack of teacher confidence and access to training, teacher capacity and the impact of the National Curriculum, inspection frameworks and exam specifications remain unchallenged (Rushton and Dunlop, 2022). In fact, the strategy only discusses the teaching of the elements of climate change and sustainability as an additional topic in geography and science. This is a missed opportunity to harness transformational change. As we will explore later in this chapter, there are many cross-curricular opportunities for not only teaching about climate change but also to develop a deeper connection to nature within children and young people.

One of the greatest criticisms of the SCCS is that the responsibility for change is soundly placed onto individual teachers and schools to incorporate this into their teaching and learning (Rushton and Dunlop, 2022). While the strategy recognises the importance of leaders to lead the way through the Sustainability Leadership initiative, this is focused on school leaders rather than policy makers. In reality, it can only be when policy makers lead the charge that effective change will occur.

We suggest that on a classroom level, and even working within a prescriptive Primary National Curriculum, space can be found to explore these important concepts. There are explicit learning objectives within both the Science and Geography sections of the curriculum that relate to the teaching of nature and the natural world (DfE, 2013), and the broad nature of the statutory requirements for the foundation subjects allows space for links to nature to be made through a cross-curricular approach. By incorporating experiential learning through a cross-curricular approach in primary schools, we can empower children with knowledge about their own planet. We can inform them about climate change, its causes and solutions. We can make opportunities for them to strengthen their connection to nature, so that they begin to care about their environment as they would care about other human beings. By doing these things we can enable them to become agents of change for their own future. By combining this approach with the Common Third, we can widen the impact to families and the wider community too.

People connectedness and the Common Third

The COVID-19 pandemic resulted in the deaths of more than 150,000 people in the UK (Office for National Statistics, 2023). During the period of March

2020 to July 2021, there were a series of national and local lockdowns in the UK, during which time teaching was provided online. While this approach supported the maintenance of academic progress, it did not allow for children to have the environmental and social experiences that would be assumed in a child's development. The impacts that this time had upon children's social interactions, communication and mental health have been widely researched, but an enduring impact in ensuing years has been pupil absence. Weale (2023) explains how this has been exacerbated by anxiety, poverty, illness and more parents working from home. The increase in school absenteeism has been noted as "staggering" by Maddern (2023), who cited research indicating that school absenteeism had doubled since 2017-2018.

With the aftermath of the pandemic still apparent and children's absence remaining an issue, the concept of people connectedness is more important than ever. Social interaction with others has been found to empower health and wellness (Martino et al., 2017). Social connection is a fundamental aspect of Maslow's Hierarchy of Needs, framed as "love and belongingness" (Maslow, 1954 cited in Martino, Pegg and Frates, 2017) and is also an integral aspect of self-determination theory, being referred to as "relatedness" by Deci and Ryan (1985, cited in Martino et al., 2017). With evidence that connecting with others is an empowering, beneficial and healthy approach, it is useful to consider how it relates to sustainability.

It is helpful to start with social isolation. This is distinct from loneliness and refers "to a lack of social connections and infrequent social interactions" (American Psychiatric Association, 2023). Social isolation can have a detrimental effect on mental health and wellbeing (Oxford Poverty and Human Development Initiative, 2024). When children are socially isolated, one of the most impactful elements is a lack of care; children feel like people do not care about them, including peers and teachers. Subsequently, they end up in a position of not caring; not caring for themselves, for others and for the environment. Therefore, it is important how we, as teachers, can help children avoid social isolation. This can be done through providing opportunities for children to connect and work together with others for a common purpose. Within education settings we have the environment and means to make a difference to how young people feel and to foster a caring, communal environment. On a basic level, committing to classrooms and learning environments that celebrate collaborative learning, group work and shared objectives will help create this. The culture within the school is also critical, with teachers working collaboratively being repeatedly noted as one of the most impactful elements on a child's progress (Zee and Koomen, 2016). If teachers establish a shared understanding of community and work together to cultivate a positive environment, this could lead to successful outcomes for all. Therefore, a school's community and culture are vital to ensure success for the children they teach, as well as developing them holistically and showing them that people care about them. This, in and of itself, will make a difference.

A social pedagogical approach to teaching focuses upon a holistic approach to children's well-being, learning and growth, with teachers working alongside young learners to enhance relationships and communication (Barton, 2021; Charfe and Gardner, 2019). Central to this approach is the concept of the Common Third. The term relates to the teacher and pupil undertaking a shared task, together working towards a mutual goal and sharing a new experience (Mygind et al., 2018, p. 606). Fundamentally, it requires minimising imbalances of power – the hierarchy between teacher and pupil is suspended for that moment in time, with the two individuals existing in a common space as equal and authentic participants (Charfe and Gardner, 2019, p. 40).

Using the Common Third approach or attempting to find that moment in time requires certain conditions. A positive and authentic relationship between the teacher and pupil is required, and being in the right physical environment also has a substantive role to play (Hadi and Johansen, 2024). The outdoor environment is often seen as a great 'leveller.' There will always be a power imbalance between teachers and children. Overcoming this requires a conscious minimisation of hierarchal thoughts or actions, working alongside the child and being open to negotiating the direction they want to take the learning. It requires discussion, compromise and understanding from both parties, toward a shared aim that has been agreed from the outset. As a teacher, this can be difficult, as teachers are used to setting goals, leading the learning and holding accountability for the progress made in that time. This approach is a shift in mindset; teachers and the child set the goal, there is no 'leader,' and the only accountability is between each other and how the activity is conducted. With the shared activity as the focus the relationship has the optimum conditions to strengthen more naturally (Charfe and Gardner, 2019).

However, it can be achievable. Forest school, outdoor adventure activities within the PE National Curriculum, residential trips and continuous provision in Early Years settings, all offer potential opportunities for the Common Third approach to teaching to be incorporated. Sometimes, this is unplanned, but what is clear is that this approach offers true connection; a positive relationship founded on mutual respect that aspires to achieve a shared goal which will benefit both the teacher and learner. This is the epitome of what it means to be connected. In the remainder of this chapter, we share examples of ways that the Common Third approach can be used to bring learners together to address issues of sustainability. The Common Third approach alongside nature connectedness, has potential to elicit transformative change for children, families and the wider community, yielding powerful results.

Exploring the Common Third in practice

Using the 5 Pathways to Connecting with Nature from the Nature Connection Handbook (Richardson and Butler, 2022) as a framework we explore some practical ideas with a cross-curricular approach. These ideas could be used as inspiration to develop children's connection or friendship with nature, each other, families and the wider community. During this section, when we use the term 'learners' we are referring to children, teachers, parents and people from the wider community who might be involved in these Common Third activities.

Senses: exploring and experiencing nature through all the senses

Subjects: history, music

The Ancient Maya is a time period teachers can choose to teach in Key Stage 2 as a non-European society which contrasts with British history (DfE, 2013). Archaeologists have discovered a variety of musical instruments from this time-period including whistles, ocarinas and flutes (Hepp *et al.*, 2014). Music was an integral part of social, religious and domestic life during this time and study of these instruments reveals musical inspiration could have been drawn from the natural world (Hepp *et al.*, 2014). During the teaching of history, replica artefacts of these instruments could be studied in the primary classroom, with children investigating their purpose and the sounds they could have made. As a cross-curricular activity with music, the children with the teachers and parents could compose their own music based on the sounds in their natural environment after a visit to an outside space. The Secret Sky Garden (Sarah and Lumbers, 2018) could be used for inspiration here. This is story of a girl, Funni, who creates a garden in a disused carpark and then, with her friend Zoo, creates music with their recorder and harmonica replicating the sounds they hear, both human-made and natural. Eventually, all learners could put the sounds together to create a composition or soundscape which then could be performed.

Beauty: seeking and appreciating the beauty of the natural world

Subjects: art, computing, poetry

The beauty of the natural world has long been an inspiration for artists, writers and poets. This has led to the creation of profound pieces of art and writing as a form of biophilia; reflecting an inherent need in humans to experience joy when they see, feel or hear something wonderous in nature (Wilson, 1984). The work of Cath Waters, an artist living and working on the Isle of Skye, could be used as a stimulus in this area. Her work is landscape-based made of photography and collage. The learners could examine the artist's work, then go into their own local environment (such as the playground, a balcony, parks, woods, rivers, lakes – wherever is nearby) and take photographs using technology such as cameras and iPads, draw pictures, collect objects, write down words and thoughts to eventually bring these back together in a shared learning space to create a mixed media collage. This could be one large collaborative piece and in doing so, create the conditions for the Common Third. Other inspirational artists that work closely with nature include Mika Ninagawa, Vanessa Hogge, Andrew Cooper, Javiera Estrada, Jakkee Kongkaew, Alexandre Manuel, Andy Goldsworthy, and Tromarama.

Emotion: noticing and welcoming the feelings nature inspires

Subjects: English, music, dance

Generations of poets and have used nature as their muse using poetry, music and dance to express their feelings. Poems such as 'Starlings in Winter' or 'The Swan' by Mary Oliver (2006; 2012) could be explored as examples of this. Learners could examine the language, style and structure Oliver uses to describe the birds as she watches them and the feelings this invokes in her. The learners could watch birds, animals or plants in their own local environment, collaboratively noting down illustrative words to describe the appearance, movements or noises of their subjects, they could also note the feelings that they experience as they watch. Such activities not only build a closer connection to nature, but also encourages learners to use and create a wider vocabulary to discuss and describe the wonder of nature and the feelings that this conjures. The structure of either poem could act as a scaffold for the learners to collaboratively write their own poetry. There is also potential to perform these using music and dance inspired by the movements and sounds of their chosen animal or plant. Other inspirational poets whose work involves a close connection to nature and the feelings it invokes include, Emily Dickinson, Fukuda Chiyo-ni (haiku), Anne Spencer and Robert Macfarlane.

Meaning: celebrating and sharing nature's events and stories

Subjects: History, English

Tapping into ancient folklore and mythology could be helpful here. Trees have often been used in ancient folktales about fairies, healing powers and warnings. For example, in Celtic mythology the Blackthorn tree was considered to have magical properties and Druids used staffs made from this tree when practising their art (Beresford-Kroeger, 2019). It was often associated with the gods of thunder, as they were very often the tallest tree in the landscape and therefore the first to be struck by lightning. Celebrating nature's resilience over the erosion caused by humankind could be explored here; how nature can reclaim what she had once lost. The Rose that Grew from Concrete by Tupac Shakur (2006) could be a stimulus here. Using the key messages of resilience and determination in the context of nature, learners could look for the ways in which nature endures despite the challenges. For example, saplings growing in cracks in the pavements, ivy growing over a derelict house, weeds reclaiming a disused playground. Learners could work together to create a similar poem or piece of artwork with the wider community.

There are many more examples of ancient folklore concerning trees here: https://www.woodlandtrust.org.uk/blog/2021/04/tree-folklore/; https://heartofenglandforest.org/news/folklore-forest

Compassion: helping and caring for nature

Subjects: Science, design and technology, Geography

This pathway is all about helping, caring for and restoring nature. It centres around learners examining what is needed for a successful habitat in science, considering how specific animals need specific habitats to survive, and what happens if these habitats are changed. The learners can then use this knowledge to create a habitat garden in the school grounds, with adults from the wider community helping to create it. This could range from a bug hotel or bird feeders in the corner of the playground to a sensory garden where the plants, shrubs and trees have been chosen to create habitats for creatures. The learners could then monitor the species of life that visit, learn their names (addressing concerns about species blindness) and research more about each new visitor, with creatures becoming a valued part of their school community.

Conclusion

We started this chapter thinking about the wanton destruction of the Sycamore Gap tree and the outrage that this caused. We suggested that despite our current society's distance from the natural world, this reaction indicated that the connection was still there deep within us. Throughout this chapter we have explored the concepts of nature connectedness and the Common Third; how the Common Third could be used as a tool to encourage a greater connection and friendship with nature and with each other. We have suggested that by combining these concepts effectively we can create transformational change in the way we see and interact with each other and with nature. We advocate that as educators, we should encourage children to see nature as a friend, and, as such, for them to consider how that friendship might be developed and strengthened. As noted in the Relationships and Health Education guidance

(DfE, 2019), "most friendships have ups and downs, and that these can often be worked through so that the friendship is repaired or even strengthened, and that resorting to violence is never right." As educators, we can empower our young children to be agents of change and form that caring friendship with nature and each other. This, we argue, could result in real change, through developing a greater understanding of sustainability and our individual and collective responses to the global ecological crisis.

References

American Psychiatric Association (2023) *Taking on the Public Health Threat of Loneliness and Social Isolation*. Available at: https://www.psychiatry.org/news-room/apa-blogs/loneliness-and-social-isolation (Accessed: 5 February 2024).

Armstrong, K. (2022) *Sacred Nature*. London: Penguin Random House.

Barton, C. (2021) 'Embedding Social Pedagogy and Psychological Safety with Virtual Teams', *International Journal of Social Pedagogy*, 10(1), pp. 1–12. Available at: https://doi.org/10.14324/111.444.ijsp.2021.v10.x.008.

Beresford-Kroeger, D. (2019) *To Speak for the Trees: My Life's Journey from Ancient Celtic Wisdom to a Healing Vision of the Forest*. London: Timber Press.

Carrington, D (2016) 'Three-Quarters of UK Children Spend Less Time Outdoors Than Prison Inmates – Survey', *The Guardian*, 25 March. Available at: https://www.theguardian.com/environment/2016/mar/25/three-quarters-of-uk-children-spend-less-time-outdoors-than-prison-inmates-survey (Accessed: 21 February 2024).

Charfe, L. and Gardner, A. (2019) *Social Pedagogy and Social Work*. London: Sage.

Department for Education (2013) *National Curriculum for England: Primary Curriculum*. Available at: https://www.gov.uk/government/publications/national-curriculum-in-england-primary-curriculum (Accessed: 27 February 2024).

Department for Education (2023) *Sustainability and Climate Change: A Strategy for the Education and Children's Services*. Available at: https://www.gov.uk/government/publications/sustainability-and-climate-change-strategy/sustainability-and-climate-change-a-strategy-for-the-education-and-childrens-services-systems (Accessed: 26 February 2024).

Flood, A. (2015) 'Oxford Junior Dictionary's Replacement of 'Natural' Words with 21st-Century Terms Sparks Outcry', *The Guardian*, 13 January. Available at: https://www.theguardian.com/books/2015/jan/13/oxford-junior-dictionary-replacement-natural-words (Accessed: 21 February 2024).

Hadi, J. and Johansen, T. (2024) 'The Common Third: A Kindred Spirit to Youth Work?' Available at: https://sppa-uk.org/the-common-third/ (Accessed: 28 February 2024).

Hepp, G., Barber, S. and Joyce, A. (2014) 'Communing with Nature, the Ancestors and the Neighbors: Ancient Ceramic Musical Instruments from Coastal Oaxaca, Mexico', *World Archaeology*, 46(3), pp. 380–399. Available at: https://doi.org/10.1080/00438243.2014.909100.

Jones, L. (2020) *Losing Eden: Why Our Minds Need the Wild*. London: Penguin Random House.

Lowry, C.A., Hollis, J.H., de Vries, A., Pan, B., Brunet, L.R., Hunt, J.R., Paton, J.F., van Kampen, E., Knight, D.M., Evans, A.K., Rook, G.A. and Lightman, S.L. (2007) 'Identification of an Immune-Responsive Mesolimbocortical Serotonergic System:

Potential Role in Regulation of Emotional Behavior', *Neuroscience*, 146(2), pp. 756–772. Available at: https://doi.org/10.1016/j.neuroscience.2007.01.067.

Maddern, K. (2023) 'New Analysis Shows "Staggering" Increase in Persistent School Absence Among Pupils', University of Exeter, 28 June. Available at: https://news.exeter.ac.uk/research/new-analysis-shows-staggering-increase-in-persistent-school-absence-among-pupils/ (Accessed: 5 February 2024).

Martino, J., Pegg, J. and Frates, E. (2017) 'The Connection Prescription: Using the Power of Social Interactions and the Deep Desire for Connectedness to Empower Health and Wellness', *American Journal of Lifestyle Medicine*, 11(6), pp. 466–475. Available at: https://www.ncbi.nlm.nih.gov/pmc/articles/PMC6125010/

Mygind, E., Bølling, M. and Barfod, K. S. (2018) 'Primary Teachers' Experiences with Weekly Education Outside the Classroom during a Year', *Education 3-13*, 47(5), pp. 599–611. Available at: https://doi.org/10.1080/03004279.2018.1513544

Natural England (2021) *The Children's People and Nature Survey (C-PANS)*. Available at: https://www.gov.uk/government/statistics/the-childrens-people-and-nature-survey-for-england-summer-holidays-2021-official-statistics/the-childrens-people-and-nature-survey-for-england-summer-holidays-2021-official-statistics (Accessed: 29 February 2024)

Office for National Statistics (2023) *Coronavirus (COVID-19) Latest Insights*. Available at: https://www.ons.gov.uk/peoplepopulationandcommunity/healthandsocialcare/conditionsanddiseases/articles/coronaviruscovid19/latestinsights (Accessed: 19 February 2024).

Oliver, M (2006) *Owls and Other Fantasies: Poems and Essays by Mary Oliver*. Boston: Beacon Press.

Oliver, M (2012) *Swan: Poems and Prose Poems*. Boston: Beacon Press.

Oxford Poverty and Human Development Initiative (2024) *Social Connectedness*. Available at: https://ophi.org.uk/research/missing-dimensions/social-connectedness/ (Accessed: 5 February 2024).

Pritchard, A., Richardson, M., Sheffield, D. and McEwan, K. (2019) 'The Relationship between Nature Connectedness and Eudaimonic Well-Being: A Meta-analysis', *Journal of Happiness Studies*, 21, pp. 1145–1167. https://doi.org/10.1007/s10902-019-00118-6.

Quing Li (2010) 'Effect of Forest Bathing on Human Immune Function', *Environmental Health and Preventative Medicine Journal*, 15, pp. 9–17. Available at: https://doi.org/10.1007/s12199-008-0068-3.

Richardson, M., Sheffield, D., Harvey, C. and Petronzi, D. (2015) *The Impact of Children's Connection to Nature: A Report for the Royal Society for the Protection of Birds (RSPB)*. University of Derby. https://doi.org/10.13140/RG.2.1.1038.5686.

Richardson, M., Hunt, A., Hinds, J., Bragg, R., Fido, D., Petronzi, D., Barbett, L., Clitherow, T. and White, M. (2019) 'A Measure of Nature Connectedness for Children and Adults: Validation, Performance, and Insights', *Sustainability*, 11(12), p. 3250. https://doi.org/10.3390/su11123250

Richardson, M. and Butler, C.W. (2022) *The Nature Connection Handbook: A Guide for Increasing People's Connection with Nature*. Available at: https://findingnatureblog.files.wordpress.com/2022/04/the-nature-connection-handbook.pdf (Accessed: 27 February 2024).

Robinson, C. (2023) *Sycamore Gap Tree: The Story So Far*. Available at: https://www.bbc.co.uk/news/uk-england-tyne-66994729 (Accessed: 5 February 2024).

Rushton, E. and Dunlop, L. (2022) The Place of Education in the Government's Draft Sustainability and Climate Change Strategy. BERA Blog [Blog post] 02.02.2022.

https://www.bera.ac.uk/blog/the-place-of-education-in-the-governments-draft-sustainability-and-climate-change-strategy.

Sarah, L. and Lumbers, F. (2018) *The Secret Sky Garden*. London: Simon and Schuster Children's UK.

Shakur, T. (2006) *The Rose That Grew from Concrete*. London: Simon and Schuster UK.

Shepherd, N. (2011) *The Living Mountain*. Edinburgh: Canongate Books.

Smith, D. G., Martinelli, R., Besra, G.S., Illarionov, P.A., Szatmari, I., Brazda, P., Allen, M.A., Xu, W., Wang, X., Nagy, L., Dowell, R.D., Rook, G.A.W., Rosa Brunet, L. and Lowry. C.A. (2019) 'Identification and Characterization of a Novel Anti-Inflammatory Lipid Isolated from Mycobacterium Vaccae, a Soil-Derived Bacterium with Immunoregulatory and Stress Resilience Properties', *Psychopharmacology*, 236(5), pp. 1653–1670. Available at: https://doi.org/10.1007/s00213-019-05253-9.

State of Nature Partnership (2023) *State of Nature Report*. Available at: https://stateofnature.org.uk/wp-content/uploads/2023/09/TP25999-State-of-Nature-main-report_2023_FULL-DOC-v12.pdf (Accessed: 21 February 2024)

Twohig-Bennett, C. and Jones, A. (2018) 'The Health Benefits of the Great Outdoors: A Systematic Review and Meta-Analysis of Greenspace Exposure and Health Outcomes', *Environmental Research Journal*, 166, pp. 628–637. Available at: https://doi.org/10.1016/j.envres.2018.06.030.

United Nations (2023) *The Sustainable Development Goals Report*. Available at: https://unstats.un.org/sdgs/report/2023/The-Sustainable-Development-Goals-Report-2023.pdf (Accessed: 26 February 2024)

University of York, UIESR and EPI (2022) *The Impact of the COVID-19 Pandemic on Children's Socioemotional Well-being and Attainment during the Reception Year*. Available at: https://educationendowmentfoundation.org.uk/projects-and-evaluation/projects/the-impact-of-the-covid-19-pandemic-on-childrens-socioemotional-well-being-and-attainment-during-the-reception-year#:~:text=Although%20most%20parents%20and%20some,pre%2Dpandemic)%20YR%20cohort%20 (Accessed: 5 February 2024).

Weale, S. (2023) 'From Covid to Poverty: Why Pupil Absence in England Is Rising', *The Guardian*, 28 June. Available at: https://www.theguardian.com/education/2023/jun/28/covid-poverty-pupil-absence-england-schools-social-economic-pandemic-families (Accessed: 5 February 2024).

Weinstein, N., Przybylski, A.K. and Ryan, R.M. (2009) 'Can Nature Make Us More Caring? Effects of Immersion in Nature on Intrinsic Aspirations and Generosity', *Personality and Social Psychology Bulletin*, 35(10), pp. 1315–1329. Available at: https://doi.org/10.1177/0146167209341649.

Wells, N.M. (2000) 'At Home with Nature: Effects of Greenness on children's Cognitive Functioning', *Environment and Behaviour*, 32(6), pp. 775–795. Available at: https://ltl.org.uk/wp-content/uploads/2019/02/at-home-with-nature.pdf.

Wilson, E. (1984) *Biophilia*. Cambridge, MA: Harvard University Press.

World Meteorological Organisation (2024) *WMO Confirms That 2023 Smashes Global Temperature Record*. Available at: https://wmo.int/media/news/wmo-confirms-2023-smashes-global-temperature-record (Accessed: 21 February 2024).

Zee, M. and Koomen, M. Y. (2016) 'Teacher Self-Efficacy and Its Effects on Classroom Processes, Student Academic Adjustment, and Teacher Well-Being: A Synthesis of 40 Years of Research', *Review of Education Research*, 86(4), pp. 985–1015. Available at: https://doi-org.apollo.worc.ac.uk/10.3102/0034654315626801.

PART 3

Leadership for positive change

Nicola Watson and Carla Solvason

In this final section we consider the intersection between leadership and sustainability. It is easy to think of leaders as being removed or separate from the majority and therefore to leave others with assumed expertise to take responsibility for 'big' environmental issues. The following chapters, however, suggest that change requires a collaborative approach. They remind us that everyone of us has a role to play in the survival of our planet. Each one is based upon 'strategic hope' – hope that is founded on action. Through collaboration positive change is 'woven' from the fabric of interpersonal and intercommunity relationships. They recognise that climate change and biodiversity loss will not be experienced equally, although every one of us will be affected. The environmental emergency transcends national boundaries and cultural barriers. By working together, we can draw upon shared understandings to deepen our own. This approach applies at every level, local, national and international.

The chapters reject the dominant idea that 'more is better' and instead consider what is 'enough' to sustain a future for ourselves and future generations. They make the case that our education systems require radical reform to embed a sense of agency in collective and environmentally just ways. They also acknowledged that educators, communities, children and young people need to be supported through this transition and provide exemplars which highlight pioneering work. Practical ideas as well as values and ideologies are shared. In the final chapter we hear how communities have come together to make positive changes with children and young people taking a lead in projects such as creating wild flower areas, bug hotels and hibernaculums, or pledging to reduce plastic waste. We are invited to consider what *we ourselves*

DOI: 10.4324/9781003506850-12

will do to make a difference and reminded that we can all ‘lead’ change in our own ways, individually and together.

The clear message of this section is that only changed thinking will bring about changed outcomes. Shared wisdom, and shared resources, will increase our capacity to ensure positive outcomes for our planet.

9

MOBILISING CAMPUS COMMUNITIES FOR CLIMATE ACTION

Student-led initiatives and collaborative approaches in universities worldwide

Gareth Dart and Ella Partington

Introduction

This chapter explores examples of student activism and participation in higher education institutions (HEIs) deemed worthy of note in academic or journalistic circles. Such noteworthiness might result from student success in shifting the practices of HEIs regarding education about climate change (CC), and/or engagement with meaningful action to mitigate the impact of CC. The examples draw upon a small sample of activities from the Global North (USA, UK, Spain, Sweden), South Asia, and sub-Saharan Africa. Examples of international networks (student to student and/or joint student-staff) are also addressed briefly. The discussion examines examples which have been directly brought about by student activism and others where there is a clear element of partnership integral to their development. Sometimes it is of course difficult to delineate precisely between these two categories, but it does illustrate the advantage of all stakeholder groups being cognisant of the variety of activities across the institution.

Contextualising the urgency of climate change within the university setting

At the time of writing the world has recorded its warmest average year on record (2023) and warmest start to a year (January 2024) (National Aeronautics and Space Administration [NASA] 2024). Ice caps are melting and sea levels rising in front of our eyes (Met Office, n.d.). The National Union of Students (National Union of Students [NUS], 2020) state that Children and Young People, who have arguably contributed least to the growing climate

DOI: 10.4324/9781003506850-13

emergency, are feeling the weight of this issue the most, and therefore 'that mounting outrage and radical action' within this generation is justified.

Although globally only 3% of the population experience higher education (HE), 80% of leaders are graduates of HEIs (Students Organising for Sustainability [SOS], n.d.). Given this presence in key decision-making roles, it is of paramount importance that students' experiences of HE prepare them to engage effectively with the variety of social, political and technological challenges that CC is unleashing. In a study of over 4,000 students across 9 HEIs in Brazil, Fiji and Kenya, Rolleston et al. (2023, p. 1) note how students entering HE have high expectations as to how their institution provides education about such matters, and how the institution itself tackles the challenges. The authors conclude, "Results indicate that satisfaction among students in the sampled HEIs is often low, while expectations are typically high." Interestingly they claim, "There is some evidence that students in contexts already more directly exposed to the impacts of climate change were somewhat more active and more satisfied" (2023, p. 1).

A recent systematic review of how 357 HEIs attempted to integrate sustainable development programmes, focusing on the period between 2013 and 2020, was pessimistic regarding many of the claims made in the research under analysis. Probst (2022, p. 2013) concluded that:

> The reviewed research was case-driven and often undertheorized regarding learning processes and outcomes. Despite its volume, the literature did not provide coherent insights into what should be learned and how.

Probst (2022) advises that HE needs to engage with more imaginative forms of education and that research on impact needs to be more rigorous in nature. There is also evidence that teaching and learning surrounding sustainable development and CC still tends toward the scientific (for example Coleman et al., 2017, Eilam et al., 2020), with broader social and political issues and notions of climate justice marginalised (Tolppanen et al., 2022).

HEIs also grapple with the contradictions of internationalisation. While working internationally fosters constructive and useful dialogue between faculty and between students, this has an environmental impact. It is also the case that many universities in the global North rely heavily on income from international students. This too incurs an environmental impact on the planet (McCowan, 2023).

Fortunately, there appears to be a growing body of evidence, some of which is explored in this chapter, that well-organised student movements, imaginative staff-student partnerships and international collaboration, can influence HEIs to work more effectively both to educate the student body about CC and to change the practices of HEIs to better reflect the nature of the emergency.

The historical development of universities and students as activists

Student activism on social and political issues is not a new phenomenon (Harmoinen et al., 2020). Well-known examples include the drive for racial equality and inclusion in the 1960's in the USA, protests against the Vietnam War, and strikes and protests against various politically oppressive regimes and poor quality HE provision in African contexts. Although there is evidence that current generations of students are less *party*-politically motivated, it would be wrong to conclude that this necessarily translates into a generally less politically motivated generation as can be seen from many instances of direct student action over the last decade. Writing from the North American context, Sloam et al. (2022, p. 22) suggest that;

> Political action is increasingly centred around everyday issues that challenge citizens' identities, and can bubble up with great speed and intensity.

Some of these movements gained international traction, but there were other, more specific student-led protests globally, including the Hong Kong protests against increasing threats to political freedom of expression and democratic processes from the Chinese government, the dramatic increase to student fees in the UK, protests in Tehran regarding the death in custody of a female protester (Reuters, 2022), protests against various Human Rights infringements in Pakistan (Amnesty International, 2020) and the demand for a decolonised HE curriculum in South Africa (Govender and Naidoo, 2023). Regarding African contexts, Zeilig and Ansell (2008, p. 31) point out that, "Any survey of postcolonial African politics reveals an astonishing level of student action that is frequently projected onto a national stage."

Cole and Heinecke (2020, p. 90) argue that student protests, though sometimes from differing perspectives, demonstrate the commitment of students to the cause of HE, even if in the form of challenge. This is because while they might have very high expectations of what their HEI can and should be offering, their visions challenge universities to imagine very different ways of being and doing (XR Universities, n.d.).

The significance of student activism in driving institutional responses to climate change

The roles and responsibilities of HEIs, staff and students can be perceived in numerous ways. UNESCO (1998) outlines the duty of HE facilities and their personnel to use their voice as a means of circulating ethical, respectful, and true views of the world, exercising their scholarly influence in a way that supports the needs and changes required for society to flourish. The action of leaders within HE is, arguably, one of the most influential factors in the change in attitudes, perspectives, and practice regarding a move towards sustainability and the fight to tackle the climate crisis (Filho et al., 2020).

We agree with Tilbury's position (Tilbury, 2011) that HEIs play an integral role in the move towards sustainable development, with such large institutions holding the power to change the world through the shaping of young minds, influencing and reshaping policy, and showing exceptional commitment to the cause through on-campus practices. This is echoed by McCowan (2021), who indicates that universities play a vital role both regionally and internationally in the move towards sustainable development and slowing the human contribution to climate change.

The levels of activism among HE staff have risen in recent years, with professors and other faculty members joining students in walk outs, strikes and cancelled classes for youth-led global climate strikes in the USA and Europe (Fisher and Nasrin, 2020). The responsibilities of leaders and faculty within HE is more complex than just their participation in strikes. We agree with McCowan's view (2021) of the need for a coordinated and well-structured framework for change, coupled with focused and intentional leadership which can result in meaningful developments surrounding these issues.

Examples of student activism

The reader might be surprised not to see a discussion of the school climate strike movement. The reason for this is that while these did overlap with HEIs they were, as the name suggests, primarily school-based and there is already insightful analysis of their nature and impact (e.g. Pickard, 2022). Instead, below, we share examples that are based firmly within HEIs.

The global North

One of the more dramatic and arguably most successful student-led campaigns in the CC arena has been the Divest from Fossil Fuel campaign (Grady-Benson and Sarathy, 2016). This started in the USA where major successes have been scored by student groups (often supported by academic staff) taking direct non-violent action against HEIs. The campaign has evolved into a series of strands involving: persuading HEIs to divest their investment portfolios from fossil fuel companies, challenging the moral appropriateness of universities acting as recruitment centres for fossil fuel companies, and challenging the ethical stance of universities that relied on such companies to support research in the HE sector.

Regarding the investment/divestment strand of the campaign, Apfel (2015) argues that what has made the campaign so successful is that it has coincided with a growing awareness in society of the harm being caused by the ongoing use of fossil fuels and a strong economic argument that continued reliance on the fossil fuel sector as a source of energy will prove to be more and more catastrophic in terms of its cost.

The divestment movement has led to further, broader, developments at the University of Barcelona where student activists occupied the university administration building demanding that it creates a compulsory course for all 14,000 students on CC alongside an education programme for all 6,000 academic staff. These programmes have now been created (with the University agreeing to the demands of the students for the involvement of particular international experts in the field in its creation) and will begin in the 2024 academic year (Burgen, 2022). The movement has now taken a step further with a campaign aimed at encouraging universities to avoid giving space to recruitment events for employers involved in the fossil fuel industry.

In 2023 students at University College London disrupted a careers fair and caused the British Petroleum stand to be shut down. The action was coordinated by the charity People & Planet which leads the project – Fossil Free Careers – across several universities. The charity claims that so far 6 universities and 15 student unions have signed up for the movement which seeks to exclude fossil fuel companies from all career activities (Pannell, 2023).

Beyond investments and careers, universities now find themselves having to take decisions on whether to divest themselves of research funding provided by the fossil fuel sector. In 2023 the Vrije University, Amsterdam, decided to pull out of all such research, a decision it admits was influenced by a series of direct actions by Dutch students across the country, including the occupation of university buildings. Writing in 'The Nation,' Cohen (2023, unpaginated) stated:

> The VU Amsterdam announcement builds on a string of FFR victories in the United States and the United Kingdom which, while still limited in scope, provide a growing foundation for greater institutional policy change.

There appears to be a broad agreement that direct student action involving high-visibility occupations and protests, well-coordinated media and social media campaigns, and a networked approach, can be effective in shifting both the dialogue and practice. What is not quite so clear is how far such effectiveness is a function of a range of other social forces such as general public awareness and acceptance of scientific realities, economic currents and moral causes.

Sub-Saharan Africa

African countries, especially West Africa, make up the majority of countries most vulnerable to the effects of climate change (Akrofi et al., 2019, Notre

Dame University, 2021). This even though a country like Niger for example contributes less than 0.1% of global harmful emissions. Africa has the largest proportion of young people for any continent but there are also reported very low levels of knowledge of the causes of climate change (West African Climate Activists at the Forefront of the Movement for Climate Justice, 2023). Despite these challenges, there are many examples of youth climate change justice activists, for example; Ezekiel Nyanfor (Liberia), Martha Agbani (Nigeria) (West African Climate Activists at the Forefront of the Movement for Climate Justice, 2023), Leah Namugerwa (Uganda), Eric Njuguna (Kenya) and Lekwa Hope Anya (Nigeria) (Nwaubani, 2022). It is interesting that apart from Lekwa Anya they appear to make little reference to the role of universities in their various outputs. Akrofi et al. (2019) note that while involvement in climate-related workshops and courses, as well as local experience of CC impact, had a positive influence on student knowledge, involvement in student-related CC clubs did not. The sample size and distribution of responses this research draws on mean that the results and conclusions need to be treated with caution.

In the Kenyan context, Nyerere et al. (2021) conclude from a review of the literature between 1999 and 2020 that while there are clear national Kenyan policies and directives relating to climate change and sustainable development, there has been very little development, with a few exceptions, of an appropriate response at Kenyan universities in terms of curriculum and teaching. The authors refer to a large survey of students by Huho (2015) which revealed that "despite the student knowledge of the climatic change and its implications, only 7.4% of the students were in agreement that climate change education should be integrated into the university curricula" (Nyerere et al., 2021, p. 10). The paper notes the formation of the Kenya Green University Network, (Environment, 2017) but doubts that it is yet making an impact. However, there are some examples across the continent of students taking explicit action within their universities to raise the climate change issue, and also of trans-national projects involving universities and students working collaboratively. For example, in South Africa where there are very high rates of climate change awareness among school students (Cambridge International, 2020).

Students' experiences in the EPIC-N project (a social enterprise that seeks to involve universities and local communities in responding to CC challenges https://www.epicn.org) appear to have a positive impact on developing student engagement with and understanding of CC. To take a couple of examples from the Africa context, six students and six staff from the University of Botswana have been working with the Gaborone City Council to seek solutions to methane produced by the city's waste disposal system. In addition, the University of Zambia, has involved academics and up to 30 students in projects with local informal settlements in the capital, Lusaka, to improve sustainability in local environments (EPIC-N, n.d.).

South Asia

Similarly to the African context, students in the South Asian region (generally considered to be Afghanistan, Bangladesh, Bhutan, India, Maldives, Nepal, Pakistan, and Sri Lanka) live in a region where the impacts of CC are already obvious and present a major disruptive threat to people's livelihoods. The cataclysmic flooding in 2022 in Pakistan attests to this. The rapid retreat of the snowline and glaciers in the mountain regions that make up the northern borders of the region threatens the reliability of the region's water supply. The monsoon that much of the area relies on is becoming chaotic (Climate Columbia, 2021) and large parts of low-lying Bangladesh, one of the most densely populated parts of the World, are simply being inundated by rising sea levels. Despite the immediacy and severity of such threats, there appears to be very little in the literature exploring the response of HEIs generally, and students in particular, to such threats.

An international survey of university students (Filho et al., 2022, p. 16) concludes that students from the South Asian region have a greater awareness than any other region of the threats caused by CC, and conjecture that "this difference is probably associated with the variety of exposure of various regions to climate risks." It is no surprise, considering the everyday reality of their rising sea levels, that Bangladeshi students appear to have the most heightened awareness. Yet this awareness appears to be poorly addressed in the country itself. In a wide-ranging and thorough analysis of the policies and practices of HEIs in the South Asian region, Mbah et al. (2022) conclude that although CC is addressed in the Bangladeshi school curricula, there is not a coordinated approach across the various levels of education and the country lacks essential understanding as to how to address the adaptions necessary to cope with the changes being wrought by CC.

Mbah et al. (2022) describe various policy responses across the region regarding CC (e.g. Pakistan will include CC education in all curricula at all levels by 2030, Sri Lanka aims to introduce education about sustainable production and consumption at all levels by 2025, and in India, the Centre for Environmental Education has as one of its goals the production and dissemination of materials regarding CC to schools and HEIs.) However, they conclude that:

> Universities and colleges in South Asia have a moral obligation to step up and develop their curriculum so that their students are prepared to act when faced with a climate-related threat but also to adapt their practices such as agricultural and infrastructural development.
>
> *(Mbah et al., 2022, p. 14)*

Although there appears to be little in the published academic or even journalistic literature on student-led initiatives to deal with CC in HEIs in the region, there are some examples of student engagement in various international partnerships, and it is to these initiatives that this chapter now turns.

Working in partnership

Linking to the South Asian discussion there is a very new initiative, the *Student Society for Climate Change Awareness* which describes its mission thus:

> ... to empower and educate youth, particularly university students in South Asia, to address climate change and climate justice.
>
> *(SSCC, 2024)*

The website (see link in the references) contains links to various courses developed for the organisation about CC and CC mitigation. There is a link to 'Campaigns' but the project is clearly in its infancy. The nature of the content and presentation of materials attest to the sponsoring organisation, the United States Department of State. It will be interesting to monitor its progress to see how successfully it engages students in the region.

An example of an organisation currently active worldwide is Students Organising for Sustainability (International). It seeks to –

> ...convene (inter)national student-led and student-focussed organisations, working on sustainability and social justice...
>
> *(SOS, n.d.)*

It aims to do this by supporting effective networks for action between HEIs and student groups, scaling up programmes for greater impact, building student leadership capacity and acting as a space for the amplification of student voice globally.

There is clear evidence of active partnership with a wide range of student and youth-based organisations, mainly in Europe and North America but with some representation also from Africa, Asia and South America. They have produced a useful international survey of student responses to CC education (though results are heavily skewed to European contexts) (SOS, 2021). Very helpfully, they produce guidance as to how student groups can use the results of the survey to leverage better CC education opportunities in their own institutions.

Another interesting student-led partnership group is the Swedish Klimatstudenterna – https://www.kth.se/en which has been effective in producing a national ranking of Swedish universities regarding their effectiveness in dealing with CC but also in persuading the Swedish ministry to take sustainability into account in the procurement procedures for HEIs. They have attracted a wide range of international student-group partners from a wide range of countries.

Beyond purely student-driven partnership organisations there are others that originate from HE institutions more broadly. It is worth highlighting the role of The Higher Education Sustainability Initiative (HESI, n.d.), which seeks to coordinate policy development, information dissemination and action between various UN entities and HEIs.

An example of a partnership at the university level is Climate-U (Climate-U, 2023) situated in University College London and working with universities in Kenya, India, Brazil, Tanzania, Indonesia and Fiji. It describes its four aims thus –

1 To support local action on climate change through the creation of participatory action research groups in universities in Brazil, Fiji, Kenya, India, Indonesia, Tanzania and the UK
2 To assess existing coverage of climate change in the curricula, research and community engagement activities of universities in the countries.
3 To contribute to theory and understanding of the impact of higher education on climate change and sustainable development.
4 To build and strengthen national, regional and global university networks and knowledge exchange on climate change (Climate-U, 2023).

The network appears to be well-developed and highly active in research across all the partners. There is an emphasis on participatory action research (PAR) with the members of the communities in which each of the partner universities is situated. The UCL team includes students from undergraduate, Masters and PhD levels. The role of PAR as an approach to engage with issues of CC is echoed in various studies (André, 2020; Wodika and Middleton, 2020).

Renewable practice

From the brief overview of various approaches to addressing CC above, some ideas can be gleaned. For student groups who need to confront authorities directly over various aspects of the HEI approach toward CC there are lessons to be learned from the success of movements that address aspects of fossil fuel divestment.

1 Build Alliances: Forming alliances with other student groups, environmental organisations, and impacted communities can amplify the impact of the advocacy efforts. Real change requires the power of as many students and staff as possible. However, it should not be assumed that there is automatic involvement from all students (Tolppanen et al., 2022).
2 Engage in purposeful Direct Action: Utilising creative tactics such as sit-ins, protests, informational campaigns, and social media activism can raise awareness and pressure decision-makers to act.

3 Foster Solidarity: Recognising the disproportionate impacts of climate change on marginalised communities (consider groups in the locality of the HEI as well as further afield) and forging alliances with frontline communities can enhance the advocacy for climate justice. Networks formed internally, regionally and internationally are vital in providing a 'strength in numbers' approach to both activism and lifestyle changes to promote sustainability. The use of digital technology and social media has become successful in networking and connecting with others committed to promoting the cause (Fisher and Nasrin, 2020).
4 Educate and Mobilise: Conduct educational workshops, events, and outreach activities to raise awareness among peers and the wider community, to mobilise support for climate action (Grady-Benson and Sarathy, 2016).

Students might also acknowledge that there is room for bottom-up approaches that incorporate sustainable and climate-friendly choices in everyday life to make the fight for change a more achievable contribution (Ojala, 2022). For example, food choices are one of the most impactful ways to incorporate the fight for climate change and sustainability into everyday life for students. This can be seen as a form of prefigurative practice, i.e. a method of influencing society sustainably through the process of reshaping habits and social norms (Raekstad and Gradin, 2020). The Climate Coalition partnered with 'Plant Based Universities' in 2023 to encourage and support Universities to transition to a 100% plant-based menu (Climate Students, n.d.), further reinforcing the idea that small, 'bottom up' changes are an effective way to tackle the growing climate emergency.

Smaller, more manageable every day and lifestyle choices can be encouraged to move towards sustainable climate change within the student body globally. Such initiatives can become the sites for curriculum-based exploration under the idea of Living Labs, explored by Filho and Ozuyar (2023). However, we should be wary of being fooled into thinking that individual action can substitute for the necessity of challenging political, economic and social norms and institutions that ingrain CC into the fabric of everyday life (Klein, 2014).

Concerning possibilities as to how HEIs, academics, students, and the wider community might work together to address CC, various ideas can be gleaned from the literature. McCowan (2023) argues that while the crisis caused by CC makes it vital for the topic to be included in the university curriculum, the gravity of the challenge should force HEIs to reconsider the way they teach and assess, such that the student experience of these processes becomes a means not just of academic learning but of crossing traditional subject boundaries, deeper critical thinking, active participation and imaginative engagement. These thoughts are echoed by Khalo and Damoah (2023) who claim that,

> ... community engagement, faculty development, campus sustainability, policy advocacy, developing student leadership, partnerships, and

> collaborations are some approaches to reviving climate change education… By implementing these approaches, universities may position themselves as hubs for climate change education, research, and action, raising a generation of conscious, involved, and equipped individuals ready to lead societal transformation in the direction of a sustainable future.

In this way the university becomes not just a preparation in advance for life post-graduation which is the way that universities are often conceptualised, but also a space where such skills are expressed in the here and now of everyday university life (Gruber et al., 2015). A space where the interaction is not just between the student and the course content but also between the various actors (students, faculty, administration, wider community) (Carrillo-Nieves et al., 2024; Cross and Congreve, 2021) which demands a greater engagement with values, skills and knowledge. To help foster such practice, Leichenko and O'Brien (2020) and André (2020) suggest integrative learning as a model to use in addressing CC. Integrative in the sense that the emotional aspects of the topic need to be acknowledged and addressed, various modes of representation need to be employed beyond the purely scientific (literature, the arts etc), and various perspectives need to be acknowledged regarding how the transformations to a more environmentally sustainable and just future can be made.

This chapter has previously touched on the role of participatory action research (PAR) when discussing the methodology of Climate-U. PAR appears in various discussions regarding a more integrated approach. For example, Nussey (2022, p. 95) argues that it can be a way of moving the debate beyond the narrow conception of the personal carbon footprint as the sole focus of action to an examination of "institutional norms, procedures and practices" and allow for engagement with groups not normally included in the debate to enable "reciprocal flows of impact and knowledge between universities and wider societies." The HEI partnership under Climate-U (2023) has many examples of PAR-focused projects and research. Regarding this notion of PAR, an observation resulting from the exploration of the literature for this chapter is how rare it is to see the work of a student or student group in the writing up of research. While many discuss the need to work more collaboratively with students or report on the outcomes of various projects and include student feedback as research participants, very few have students as co-producers of outputs.

Overcoming challenges and achieving progress

Numerous challenges confront those who wish for HEIs to take an increasingly responsible approach to issues of CC. The legislative environment in which protest occurs is becoming increasingly hostile in many countries. For example, Ugandan climate protesters spent time in a maximum-security prison (*PressReader.com - Digital Newspaper & Magazine Subscriptions*, 2025), and there are tighter restrictions in the UK as to what constitutes a reasonable level of protest and where that can occur. In certain states of the USA, there is increasing political control over what universities can teach. A study of college textbooks in the US (Ansari and Landin, 2022) demonstrates that in the 2010s coverage of CC actually diminished. Solutions to CC reduced from 15% of the content in the 1990s to 3%. Even within HEIs there are still challenges at the curriculum level. For example, a recent study in the UK and Ireland of geography courses at HEIs revealed that CC was still not being framed as a 'wicked problem and thus potentially underplayed as an issue of concern' (Cole and Heinecke, 2020, p. 91).

Apart from the Divest from Fossil Fuel campaigns, research for this chapter seems to indicate that there is surprisingly little systematic analysis of other forms of student protest or the impact of various other partnership-style approaches to CC between students and the HEI community. As noted above, the voice of the student as co-researcher is very rare.

Conclusion

This chapter briefly sketched out the (sometimes vexed) position that HEIs find themselves in concerning CC. After an historical overview of student protest, it highlighted responses to CC from a small sample of HEIs across the global north, sub-Saharan Africa, and South Asia. There is evidence of dissatisfaction from students with how HEIs have responded to the challenge of CC, both in terms of their operations and the curriculum offer. This dissatisfaction has manifested itself in several radical protests over the last decade which have helped shift the dialogue and practice in a number of universities. Broad lessons for student groups finding themselves in disagreement with their institutions are briefly discussed.

Conflict between student groups and HEIs is not inevitable. There are good examples of the two sides working in partnership. Many HEIs are keen to move toward practices which provide their students with an educational experience that better equips them with requisite skills and knowledge to face the challenges of CC. Broadly speaking, policy frameworks internationally encourage such developments, but this can sometimes conflict with shorter-term financial imperatives and other political trends. There are useful examples of associations that bring together student activism, and others that foster collaborative working within and between universities, students, academics, local communities, and policymakers. We agree the need for greater integrated learning, including assessment practices, and wider use of PAR, to more effectively engage the HEI community in addressing the issues that attend global climate change.

References

Akrofi, M., Antwi, S. and Gumbo, J. (2019) 'Students in Climate Action: A Study of Some Influential Factors and Implications of Knowledge Gaps in Africa', *Environments*, 6(2), p. 12. Available at: https://doi.org/10.3390/environments6020012.

Amnesty International (2020) *Pakistan: The Disappeared of Balochistan*. Available at: https://www.amnesty.org/en/documents/asa33/3334/2020/en/.

André, R. (2020) 'Teaching Climate Leadership: Promoting Integrative Learning in Courses on Strong Sustainability', *Journal of Management Education*, 44(6), pp. 766–793. Available at: https://doi.org/10.1177/1052562920941547.

Ansari, A. and Landin, J. (2022) *Coverage of Climate Change in Introductory Biology Textbooks, 1970–2019*, ProQuest, p. e0278532. Available at: https://doi.org/10.1371/journal.pone.0278532.

Apfel, D.C. (2015) 'Exploring Divestment as a Strategy for Change: An Evaluation of the History, Success, and Challenges of Fossil Fuel Divestment', *Social Research*, 82(4), pp. 913–937. Available at: https://www.jstor.org/stable/44282147.

Burgen, S. (2022) *Barcelona Students to Take Mandatory Climate Crisis Module from 2024*. Available at: https://www.theguardian.com/world/2022/nov/12/barcelona-students-to-take-mandatory-climate-crisis-module-from-2024.

Cambridge International (2020) *South African Students Say Climate Change Is the Biggest Issue Facing the World Today*. [online]. Available at: https://www.cambridgeinternational.org/news/news-details/view/south-african-students-say-climate-change-is-the-biggest-issue-facing-the-world-today-20200302/ (Accessed: 20 February 2024).

Carrillo-Nieves, D., Clarke-Crespo, E., Cervantes-Aviles, P., Cuevas-Cancino, M, and Vanoye-Garcis, A. Y. (2024) 'Designing Learning Experiences on Climate Change for Undergraduate Students of Different Majors', *Frontiers in Education*, p. 9. Available at: https://doi.org/10.3389/feduc.2024.1284593.

Climate Columbia (2021) *Climate Change Is Making Indian Monsoon Seasons More Chaotic*, State of the Planet. Available at: https://news.climate.columbia.edu/2021/04/14/climate-change-indian-monsoon/.

Climate Students (n.d.) *Climate Students Movement*. Available at: https://climatestudents.com/framework/ (Accessed: 4 May 2023).

Climate -U (2023). *Five Conditions for Participatory Action Research to Enhance Universities' Contributions to Climate Justice*. Transforming Universities for a Changing Climate, Working Paper Series No. 19.

Cohen, I. (2023) *A Dutch University Just Set a Powerful Precedent for Climate Research*. www.thenation.com. Available at: https://www.thenation.com/article/environment/fossil-free-research-policy-vrije-universiteit-amsterdam-climate-change/ (Accessed: 29 February 2024).

Cole, R.M. and Heinecke, W.F. (2020) 'Higher Education after Neoliberalism: Student Activism as a Guiding Light', *Policy Futures in Education*, 18(1), pp. 90–116. Available at: https://doi.org/10.1177/1478210318767459.

Coleman, K., Murdoch, J., Rayback, S., Seidl, A. and Wallin, K. (2017) 'Students' Understanding of Sustainability and Climate Change Across Linked Service-Learning Courses', *Journal of Geoscience Education*, 65(2), pp. 158–167. Available at: https://doi.org/10.5408/16-168.1.

Cross, I.D. and Congreve, A. (2021) 'Teaching (Super) Wicked Problems: Authentic Learning about Climate Change', *Journal of Geography in Higher Education*, 45(4), pp. 491–516. Available at: https://doi.org/10.1080/03098265.2020.1849066.

Eilam, E., Prasad, V. and Widdop Quinton, H. (2020) 'Climate Change Education: Mapping the Nature of Climate Change, the Content Knowledge and Examination of Enactment in Upper Secondary Victorian Curriculum', *Sustainability*, 12(2), online. https://doi.org/10.3390/su12020591.

Environment, U. (2017) *Kenya Green University Network, UNEP - UN Environment Programme*. Available at: https://www.unep.org/explore-topics/education-environment/why-does-education-and-environment-matter/green-university-0 (Accessed: 3 March 2025).

EPIC (n.d.) *EPIC-Africa – EPIC-N*. [online]. Epic-Africa. Available at: https://www.epicn.org/epic-africa/ (Accessed: 24 February 2024).

Filho, W.L. et al. (2020) 'Sustainability Leadership in Higher Education Institutions: An Overview of Challenges', *Sustainability*, 12(9), p. 3761. Available at: https://www.mdpi.com/2071-1050/12/9/3761.

Filho, W.L. et al. (2022) *Attitudes and Perceptions of University Students on Climate Change: Evidences from an International Study*, preprint. In Review. Available at: https://doi.org/10.21203/rs.3.rs-1179074/v1.

Filho, W.L. and Ozuyar, P.G. (2023) 'Living Labs in the Context of the UN Sustainable Development Goals: State of the Art', *Sustainability Science*, 18(3), pp. 1163–1179. Available at: https://doi.org/10.1007/s11625-022-01240-w.

Fisher, D.R. and Nasrin, S. (2020) 'Climate Activism and Its Effects', *WIREs Climate Change*, 12(1). Available at: https://doi.org/10.1002/wcc.683.

Govender, L. and Naidal, D. (2023) 'Decolonial Insights for Transforming the Higher Education Curriculum in South Africa', *Curriculum Perspectives*, 43 (Suppl. 1), pp. 59–71 https://doi.org/10.1007/s41297-023-00200-3.

Grady-Benson, J. and Sarathy, B. (2016) 'Fossil Fuel Divestment in US Higher Education: Student-Led Organising for Climate Justice', *Local Environment*, 21(6), pp. 661–681. Available at: https://doi.org/10.1080/13549839.2015.1009825.

Gruber, J.S. et al. (2015) 'Enhancing Climate Change Adaptation: Strategies for Community Engagement and University-Community Partnerships', *Journal of Environmental Studies and Sciences*, 7(1), pp. 10–24. Available at: https://doi.org/10.1007/s13412-015-0232-1.

Harmoinen, S., Koivu, K. and Pääsky, L. (2020) 'University Students' Readiness for Social Activity in Climate Actions', *Discourse and Communication for Sustainable Education*, 11(1), pp. 134–152. Available at: https://doi.org/10.2478/dcse-2020-0012

HESI (n.d.) *Higher Education Sustainability Initiative* | Department of Economic and Social Affairs, sdgs.un.org. Available at: https://sdgs.un.org/HESI.

Huho, J. (2015) 'Climate Change Knowledge Gap in Education System in Kenya', *International Journal of Innovation and Research in Educational Sciences*, 2(3), pp. 2349–5219.

Khalo, X. and Damoah, B. (2023) 'Reinvigorating Climate Change Education in Universities a Social Transformative Agenda', *Environmental Science & Sustainable Development*, 8(4), pp. 19–26. https://doi.org/10.21625/essd.v8i4.1013.

Klein, N. (2014) *This Changes Everything: Capitalism vs the Climate*. London: Penguin Books Ltd.

Lakhani, N. (2024) Ugandan Climate Activists Face Charges after a Month in Maximum Security Jail. *The Guardian*, 6 February. Available at: https://www.theguardian.com/world/2024/feb/06/uganda-climate-pipeline-activists-students-jail-charges
Leichenko, R. and O'Brien, K. (2020) 'Teaching Climate Change in the Anthropocene: An Integrative Approach', *Anthropocene*, 30, p. 100241. Available at: https://doi.org/10.1016/j.ancene.2020.100241.
Mbah, M.F., Shingruf, A. and Molthan-Hill, P. (2022) 'Policies and Practices of Climate Change Education in South Asia: Towards a Support Framework for an impactful Climate Change Adaptation', *Climate Action*, 1(1), p. 28. Available at: https://doi.org/10.1007/s44168-022-00028-z.
McCowan, T. (2021) *The Impact of Universities on Climate Change: A Theoretical Framework*. Transforming Universities for a Changing Climate, Working Paper Series No. 1. Climate-U, UCL (Accessed: 4 February 2024).
McCowan, T. (2023) 'The Climate Crisis as a Driver for Pedagogical Renewal in Higher Education', *Teaching in Higher Education*, 28(5), pp. 933–952.
Met Office (n.d.) *Past and Future Sea Level Rise*, Met Office. Available at: https://www.metoffice.gov.uk/weather/climate-change/organisations-and-reports/past-and-future-sea-level-rise.
National Aeronautics and Space Administration (NASA) - Five Factors to Explain the Record Heat in 2023 (2024) earthobservatory.nasa.gov. Available at: https://earthobservatory.nasa.gov/images/152313/five-factors-to-explain-the-record-heat-in-2023.
National Union for Students (NUS) (2020) *Declaring a Climate Emergency. Green New Deal for FE and HE*. Available at: https://www.nus.org.uk/climate#_Toc45715443 (Accessed: 23 March 2024).
Notre Dame University (2021) *Vulnerability Rankings | ND-GAIN Index (2021)* gain-new.crc.nd.edu. Available at: https://gain-new.crc.nd.edu/ranking/vulnerability.
Nussey, C. (2022) 'Building University Capabilities to Respond to Climate Change through Participatory Action Research: Towards a Comparative Analytical Framework', *Journal of Human Development and Capabilities*, 23(1), pp. 95–115, https://doi.org/10.1080/19452829.2021.2014427.
Nwaubani, A.T. (2022) 'Young African Climate Activists Demand Real Change at COP27', *Reuters*, 11 November. Available at: https://www.reuters.com/business/cop/young-african-climate-activists-demand-real-change-cop27-2022-11-11/.
Nyerere, J., Gatwiri, W. and Okinyi, R. (2021) *Kenya's Climate Change Policy Actions and the Response of Higher Education*. Transforming Universities for a Changing Climate, Working Paper Series No. 4.
Ojala, M. (2022) 'Prefiguring Sustainable Futures? Young People's Strategies to Deal with Conflicts about Climate-Friendly Food Choices and Implications for Transformative Learning', *Environmental Education Research*, 28(8), pp. 1157–1174.
Pannell, A. (2023) *UCL Student Climate Change Protesters Disrupt BP Recruitment Event, UCL*. Available at: https://thetab.com/uk/london/2023/10/25/ucl-student-climate-change-protesters-disrupt-bp-recruitment-event-51354 (Accessed: 29 January 2024).
Pickard, S. (2022) 'Young Environmental Activists and Do-It-Ourselves (DIO) Politics: Collective Engagement, Generational Agency, Efficacy, Belonging and Hope', *Journal of Youth Studies*, 25(6), pp. 730–750. https://doi.org/10.1080/13676261.2022.2046258

PressReader.com - Digital Newspaper & Magazine Subscriptions (2025) Pressreader. com. PressReader. Available at: https://www.pressreader.com/usa/the-guardian-usa/20240207/282282440199787 (Accessed: 3 March 2025)

Probst, L. (2022) 'Higher Education for Sustainability: A Critical Review of the Empirical Evidence 2013–2020', *Sustainability*, 14(6), p. 3402. Available at: https://doi.org/10.3390/su14063402.

Raekstad, P., and Gradin, S.S. (2020) *Prefigurative Politics: Building Tomorrow Today*. Cambridge: Polity Press.

Reuters (2022) Iran Protests: Riot Police Use Teargas on Students at Sharif University. *The Guardian*. Available at: https://www.theguardian.com/world/2022/oct/03/iran-protests-riot-police-clash-with-students-at-sharif-university.

Rolleston, C. et al. (2023) 'Aiming Higher? Implications for Higher Education of Students' Views on Education for Climate Justice', *Sustainability*, 15(19), p. 14473. Available at: https://doi.org/10.3390/su151914473.

Sloam, J., Pickard, S. and Henn, M. (2022) 'Young People and Environmental Activism: The Transformation of Democratic Politics', *Journal of Youth Studies*, 25(6), pp. 1–9. Available at: https://doi.org/10.1080/13676261.2022.2056678.

SOS (n.d.) *Students Organizing for Sustainability International*. Available at: https://sos.earth/about/ (Accessed: 18 November 2023).

SOS (2021) *Students, Sustainability and Education*. Available at: https://sos.earth/wp-content/uploads/2021/02/SOS-International-Sustainability-in-Education-International-Survey-Report_FINAL.pdf.

SSCC (2024) *Student Society for Climate Change Awareness*. Available at: https://sscc2024.com/ (Accessed: 20 November 2023).

Tilbury, D. (2011) 'Higher Education for Sustainability: A Global Overview of Commitment and Progress', *Higher Education in the World*, 4(1), pp. 18–28.

Tolppanen, S., Kang, J. and Riuttanen, L. (2022) 'Changes in Students' Knowledge, Values, Worldview, and Willingness to Take Mitigative Climate Action after Attending a Course on Holistic Climate Change Education', *Journal of Cleaner Production*, 373, p. 133865. Available at: https://doi.org/10.1016/j.jclepro.2022.133865.

United Nations Education, Scientific and Cultural Organisation (UNESCO) (1998) *World Declaration on Higher Education for the Twenty-First Century: Vision and Action*. Available at: https://unesdoc.unesco.org/ark:/48223/pf0000141952 (Accessed: 23 March 2023)

West African Climate Activists at the Forefront of the Movement for Climate Justice (2023) The Climate Reality Project. Available at: https://www.climaterealityproject.org/blog/west-african-climate-activists-forefront-movement-climate-justice.

Wodika, A.B. and Middleton, W.K. (2020) 'Climate Change Advocacy: Exploring Links between Student Empowerment and Civic Engagement', *International Journal of Sustainability in Higher Education*, 21(6), pp. 1209–1231. Available at: https://doi.org/10.1108/IJSHE-03-2020-0091.

XR Universities (n.d.) *XR Universities' Declaration*, Google Docs. Available at: https://docs.google.com/document/d/1HL4sbxdfkBnqsDbh1-yLOB5o-0OLjLzG-ZTZzobcOMg8/edit (Accessed: 20 November 2023).

Zeilig, L. and Ansell, N. (2008) 'Spaces and Scales of African Student Activism: Senegalese and Zimbabwean University Students at the Intersection of Campus, Nation and Globe', *Antipode*, 40(1), pp. 31–54. Available at: https://doi.org/10.1111/j.1467-8330.2008.00570.x.

10

SUSTAINABILITY LEADERSHIP

A praxis of hope

Shaun McInerney and Tamzin Ractliffe

Introduction

Despite the stark reality that we may be the first species in the history of the planet to be responsible for its own extinction, we humans seem unable to lead ourselves out of the existential predicament we have created. The reasons for this are manifold and complex. Bryan (2021) notes that the inconceivable, imperceptible, and unintelligible nature of the problem combined with the fact that we are embedded in – and perpetuating of – the harmful systems at the root of the issue, leaves us entangled in patterns of denial. However, Laloux (2016, p. 6) suggests, "there is reason to be deeply hopeful. The pain we feel is the pain of something old that is dying.... while something new is waiting to be born."

How might we interrupt the bleakness of this pattern to weave a different response and offer hope to those whose lives we affect? Not the hope that is naive or perpetuating of our denial, but hope that Homer-Dixon (2021) describes as realistic, astute, and powerful. We call it strategic hope: hope that is grounded in action, strategy, and the belief in the capacity for change. Strategic hope involves a deliberate choice to focus on possibilities for positive change and the roles individuals and communities can play in creating a resilient future.

This chapter explores the potential for sustainability leadership in education to offer a frame of reference for a new and different response to the ecological crisis we face. It is written for educators who feel compelled to develop their own 'praxis' to offer alternatives to the young people whose future aspirations they influence and shape. Praxis is 'reflection on, and in, human action' (Freire and Ramos, 1970). Praxis can be seen as a gateway to sustainability leadership because it affords us the opportunity to change the

DOI: 10.4324/9781003506850-14

world through what we do, as we do it. We outline how 'praxis' can be a pathway to a more regenerative approach as educators.

We unpack two key challenges that are shaping our context as educators. Firstly, the multiple and overlapping crises we face – the 'polycrisis.' Secondly, the way we have been 'taught' to think. A sustainable approach relies on us taking conscious charge of our thinking processes and helping others do the same (Sanford, 2020, p. 1). As Einstein purportedly said, we cannot solve our problems from the same level of thinking that created them.

We propose sustainability leadership as a 'hope-full' alternative, one that might help us think differently so that we can act differently. We point to emerging practice in leadership development in the English system, some informed by adult developmental theory, which offers a way forward as we develop the capacities and shape the conditions that will enable us to develop new capabilities that support our individual and collective journeys towards a more sustainable world.

It is our hope that this chapter will offer an orientation to sustainability leadership, inspire a desire to do more to develop this capability and signpost some resources that support this. Above all, it is a call to action. We write from our own sense of hope: that humanity realises its potential to be creators and custodians of a sustainable and equitable future for all.

The challenge of our contemporary context: the global polycrisis

A recent assessment of the climate (IPCC, 2023) reflects global warming and environmental change happening across the planet at an accelerated pace. The rapidity of change is increasingly challenging the limits of scientific understanding of the climate system. Our sustainability as a species is increasingly questionable. Without radical and transformative change, many believe we are facing a catastrophic and existential threat to humanity.

This is known as the time of the polycrisis, a time in which we don't face multiple isolated crises but rather multiple, cascading and interconnected – or *entangled* – crises Lawrence et al. (2022). These create compounded impacts which lead to greater collective harm than isolated crises. Churchman (1967), an American philosopher, identifies the polycrisis as a 'wicked problem,' one that has undefined boundaries, is difficult to intervene in and will vary significantly across regions with diverse experiences and interpretations. This means that it is impossible to 'solve' in any traditional sense. We must find ways to learn to live with the 'baked in' consequences and identify ways to navigate through it towards a different way of living in the world. This is the essence of sustainability leadership.

The hope of sustainability leadership

Hope is a quintessentially human disposition that shapes our view of the world. Hope holds the tension between our human need for both self-preservation (fear) and our desire for self-transformation and flourishing (love). In this sense, hope allows us to enact a choice; it defines our limits and possibilities (Rowson, 2017). From the vantage point of the 'polycrisis' hope seems elusive at best – at worst, an indulgent displacement activity that distracts while the structures of collapse remain in place. However, operating on the level of hope can engender a deeper connection with ourselves and with one another. Hope sustains love; it emanates from our deepest sense of both being (who we are) and knowing (how we experience the world).

Leadership has a choice: To take a deficit-based approach that induces fear; or an asset-based approach that fortifies a sense of agency and inspires collective action. When we take an asset-based approach, hope is a source of energy and a resource for action. Building from this place of hope allows us to challenge the worldviews, institutions, and technologies of the old order by replacing them with ones that can better serve us and the planet (Homer-Dixon, 2021).

Sustainability leadership offers an alternative paradigm that invites us ALL to be powerful in the face of the challenges we face; to give ourselves permission to make promises beyond our capabilities and to commit to grow ourselves into skilful action. This hope recognises that we will, by necessity, evolve a next wave of principles, values, social and economic relations as we gradually adjust to the reality of a warming planet, denuded soil, biodiversity loss and warming oceans and the negative externalities of our failure to deal with this sooner.

What does sustainability mean?

For the purposes of this discussion, we propose defining sustainability as "*enough, for all, to thrive, forever*" (unknown attribution). Our reasoning for this choice is that it integrates core principles including:

The principle of enough

This term emphasises the importance of meeting the needs of all people in ways that avoid overconsumption and waste. By advocating for "enough," this definition implies not just sufficiency but also the prudent and mindful use of resources in a way that anticipates future needs and challenges. It suggests building systems that are efficient and adaptable, capable of maintaining balance and sufficiency through fluctuations and shifts in environmental conditions and resource availability. This inherently calls for resilient practices that can adjust to ensure continued sufficiency.

Inclusivity and equity

Sustainability must consider all individuals and communities, promoting fair access to resources, opportunities, and rights. It underscores the social and environmental justice dimensions of sustainability, emphasising that economic and environmental policies should benefit everyone, not just a few, and implicitly calling for socially resilient frameworks that adapt to changing dynamics.

Resilience and regeneration

The notion of forever incorporates the idea of regeneration and long-term thinking as well as the resilience and adaptive capacity of systems – whether social, ecological, or economic. This means sustainability must be conceived in a way that not only protects resources for the current generation but that also supports their renewal and availability for future generations. This implies a commitment to practices that restore and rejuvenate the natural environment, ensuring its continued vitality and productivity. It means anticipating and being able to withstand temporal changes, including environmental shifts, technological advancements, and evolving societal needs.

Sustainability leadership: a new educational paradigm

Sustainability leadership affords the possibility of leadership for a better world. Polly Courtice, Founding Director of the Cambridge Institute for Sustainability Leadership (CISL) once commented

> A leader is someone who crafts a vision and inspires people to act collectively to make it happen, responding to whatever changes and challenges arise along the way. A sustainability leader is someone who inspires and supports action towards a better world.
>
> *(pers. comm, 2017)*

Drawing on complexity science, Ferdig (2007, p. 27) outlines three core assumptions of sustainability leadership that emphasise it's about all of us, collaborating with a systems awareness: (1) anyone can choose to become 'A Leader' and take responsibility for fostering sustainable conditions in workplaces, communities and even on a global scale; (2) the role of a leader includes capabilities beyond those we currently attribute to leaders, primarily, learning what it means to be a leader 'with' others instead of a leader 'of' or 'over' others; (3) a leader cannot effectively operate outside of the holistic interconnections that exist among and between people and natural systems.

The need for a paradigm shift towards sustainability leadership in education prompted UNESCO in their International Commission on the Futures of Education report (2021) to call for a new social contract for education, emphasising that "in an era of rapid change and shifting societal needs, it is imperative to re-evaluate the principles and foundations that underpin education." A paradigm shift is a fundamental change in approach built from a different set of assumptions based on how the world is changing.

What is the work we need education to do?

Educational institutions remain one of – if not the – most important socialisation systems for the development of citizens and leaders. It is education that fundamentally informs and influences the mindset, capacities, and capabilities we develop to respond to our global context. The reality is that education systems are not evolving to allow young people to respond to a changing world. At this critical juncture in human history a reimagined educational paradigm with visionary leadership is essential to equip individuals with the values and means to navigate and mitigate our eco-predicament and drive systemic change.

Sustainability leadership embodies many 'new paradigm' values and can help educators, schools and ultimately, young people respond to a changing world. The Table 10.1 characterises some of the shifts that international non-governmental organisation, Ashoka, see emerging in the new paradigm.

TABLE 10.1 Old and new paradigms

Old Paradigm	*a. New Paradigm*
Efficiency in repetition	Defined by change
Fixed hierarchy	Shared leadership
Top-down	Distributed power
Repeat one set of skills	People-centred purpose-driven
Silos	Continuous learning
Transactions	Collaboration & co-creation
Limited access to knowledge	Relationships
Short-termism	Open and transparent

The implications of this for education systems are clear. Reframing our perspective to encompass the wider systems of which we are a part is one of the most significant challenges of the current education system. Current education models, instead, are more focused on transmissive learning that ignores the appreciation of nested system within which we are all embedded (Sterling, 2001, p. 51).

In England, the Climate Change and Sustainability strategy (Department for Education, 2023) required all schools to have a sustainability leader in place by 2025. The strategy makes clear what tasks need to be done. It does not expand on how to do these. The rationale for *why* we need to respond to a changing world is widely accepted. We are suggesting that *what* we need to do, and *how* we might do it, will emerge if we create conditions and develop capacities that unlock the latent potential in educators to adopt a sustainability lens in their schools and wider ecosystem.

Sustainability leadership is the pathway through which we might reimagine a model of education to meet these challenges. This would embed core principles across all levels of self, school, and society for all ages including:

Systems and complexity

Understanding that what we *know* arises through interaction, negotiation, and collaborative activity in the 'real world' means we must find a way of transforming lived experience into valuable capabilities. Making use of place-based eco-pedagogies – lived experience and engagement in natural and cultural contexts – is essential for co-constructing knowledge of how we are connected, interdependent, and intertwined with the non-human elements of our world. This encourages the development of ecological literacy, developing our appreciation of ourselves as part of a larger whole, where the value code is that which increases the civil commons (Howard, 2018, p. 9).

Interdisciplinary learning

Engaging with multiple perspectives on any concept is key to developing a sense and an experience of complexity. To navigate the complexity of modern challenges, education must incorporate a broad intellectual foundation, integrating insights from various fields and developing the capacity to synthesise diverse perspectives and apply them effectively and innovatively.

Agility and adaptive thinking

The ability to respond, iterate, adapt, and change, is core to the effective application of knowing into doing. This means we must be able to dive deeply into specifics without losing sight of the bigger picture.

Collaborative co-creative and participatory learning

In a future that requires us to navigate and manage complexity and discontinuity, the importance of communication, cooperation, and collaboration between diverse others cannot be underestimated. It means we must be able to align our common goals, leverage our unique strengths and optimise collective intelligence. A mindset of 'we' must predominate over 'me.' As importantly, we must view our students as participants and collaborators in their learning journey.

Practical skills

Knowing 'how to' is a neglected aspect of contemporary education that requires practical, critical, and problem-solving skills as well as a commitment to lifelong learning. Understanding how to harness the benefits – and minimise the harms – of modern methodologies like data science and design thinking are essential tools for tackling contemporary issues, making informed decisions, and driving innovation.

Is our thinking up to the challenge?

A significant challenge for education – and for leadership generally – stems from the speed of modern cultural and technological advances. Rees (2023) proposes that the rapid evolution of the socio-technological environment has outpaced the slower biological evolution of the human brain, rendering its inherited cognitive processes functionally obsolete for dealing with today's complex global challenges. As the pace of change continues to accelerate beyond the capabilities of traditional approaches to education, there is an increasing disconnect between what skills exist, what skills are being offered and what skills are and will be needed in the face of the polycrisis. Functional obsolescence reinforces the necessity for a paradigm shift in how we are prepared, both as educators and citizens, to navigate this complex landscape. New capabilities are needed to guide a different course of action: it is only changed thinking that will change outcomes.

What follows are some ideas to expand our practice. The first step in sustainability leadership is to be able to question our current reality so that we can see the paradigms within which we operate and the extent to which they serve us. This is important because we need young people to be able to do this too. We spend much more of our time than we realise making decisions intuitively rather than rationally (Kahneman, 2011). This leaves us susceptible to cognitive biases that perpetuate our tendency to resist change. Changing this requires a transformational approach to education. This has to start with ourselves as sustainability leaders – in whatever role we have – reflecting on how we need to develop ourselves into new ways of thinking and doing.

From activism to praxis

Sustainability leadership is a call to action that goes beyond activism. It can be thought of as embodied praxis. It is embodied in the sense that it integrates our emotional world, our cognition, and our practical action.

Sustainability leadership is grounded in the everyday opportunities that teachers have to sow seeds of change through their daily activities. It seeks to engage us all in collective social and societal action (Lawson and Read, 2022). In practical terms this means expanding our worldviews as educators and as global citizens and in our local contexts to foster sustainable conditions wherever we are. Being intentional about this is 'praxis.' Being conscious of the impacts of our praxis expands our own journey of reflection and our capacity for leadership. Praxis relies on reflexive thinking which means being open to questioning our basic assumptions. This invites us to be courageous: as we reflect deeply about ourselves and our work, we may experience an unsettling, or "an insecurity regarding the discourse and practices used in describing reality" (Pollner, 1991, p. 370). This reflexive approach is necessary however, because it allows us to see ourselves within the wider systems we inhabit – as teachers, parents, consumers and leaders as we ask questions about who we are, how we relate to others and what we do to take ethical action in service of a more equitable and sustainable future. This represents a 'bottom up' approach to leadership; one that does not invest authority in others to make decisions but instead encourages us to take responsibility within our scope of influence, to connect with others and build greater, shared capacities for change. This does not negate the need for wider policy and systems change but lays out a pathway of hope towards it.

The potential of adult development

Students of child development will be familiar with the work of Jean Piaget. It is interesting for educators that while we often apply Piaget's insights to the children we teach; we rarely consider their implications for our own growth as adults. Rowson (2017) reminds us that Piaget's genius lies in revealing not only *what* we know but also *how* we structure our ways of knowing. This is significant because it shapes how we understand ourselves, our ways of being and underscores our continual growth, or 'becoming.' By intentionally focusing on our own development and maturity as adults, we might develop and model sustainable leadership behaviours, such as empathy, cultivate higher-order, nuanced thinking, and adopt perspectives guided by a positive intent for others and the common good.

Over decades researchers have considered how this process of growth and development might apply to adults through the frame of needs (Maslow, 1951); how our personality, ego and capacities develop (Kegan, 1982, 1998; Hy and Loevinger, 2014) and the potential of this for intentional development of individual and collective consciousness (Graves, 1974; Wilber, 2000). For Cook-Greuter (2004) this development happens on an imaginary vertical plane. In contrast to horizontal development that sees us get better at what we are already doing, vertical development refers to "how we learn to see the world through new eyes, how we change our interpretations of experience and how we transform our views of reality" (p. 276). This has implications for professional learning which tends to adopt a task focused, in contrast to a people-focused developmental approach that supports us to evolve and grow and therefore become more sophisticated in our roles.

The work of vertical development (and sustainability leadership) is to become increasingly aware of our own selves and our mental patterns. As we become more aware of our own perspectives (why we think and act how we do, our strengths and limitations) we can appreciate others more fully for who they are and why they hold the views they do. As we become more nuanced in our perspective-taking we can accept that others, from their own level of development, may not be able to see what we see, yet. Our ability to depersonalise this, to separate who we are speaking with (the subject) from what they are saying (object) marks a shift in perspective that affords greater, and deeper understanding (Watkins, 2015, p. 201).

As we grow and expand our worldview, we are not leaving behind who we are or what we thought; but integrating it – allowing it to co-exist with our previous perspectives. We become less egocentric as we develop a worldview which is more complex, dynamic, and world/socio-centric (Cook-Greuter in Gilbride et al., 2020, p. 235). As we become more aware of others, we can appreciate that our maturity is dependent on them. Our understanding of the world is shaped by theirs and vice versa. Understanding this 'inter-subjectivity' recognises that we become who we are in relationship to, and through our interactions with others. On an interpersonal level this helps us appreciate the importance of relationships for learning and growth. In Southern Africa this is called Ubuntu: (I am because you are). At a system level it suggests that our own growth and maturity is dependent on the growth and maturity of others. In the struggle for a more sustainable world, we are in this together.

Critics of social change strategies that place the onus of responsibility on the individual argue it is the structures of society that enable or prevent change (Chater and George, 2023). Rather than dwelling in binary either/or thinking, increasing our maturity helps us engage 'yes/and' thinking. Taking a 'yes/and' approach to this challenge starts with appreciating this viewpoint, recognising that there are structures of power, deeply entrenched worldviews, and transmission mechanisms (such as the education system) that perpetuate the status quo. And, at the same time, building the social demand for systemic change at the individual and relational level gives politicians and policy makers the rationale to act.

How does one become a sustainability leader?

Sustainability leadership has much in common with 'changemaking' – a leadership approach pioneered by international NGO, Ashoka. Changemaking emphasises four skills: the importance of acting from empathy; the capacity to work in non-siloed, diverse and multidisciplinary 'teams of teams' to solve complex, wicked problems; the shared and collective responsibility to step up and contribute according to our own capacity and the action of 'practising changemaking' through which we grow through reflexive thinking and action, or praxis.

> Sustainability leaders make the notion of 'sustainability' personally relevant, grounding action in a personal ethic that reaches beyond self-interest. They recognize that all of us can co-create the future through individual ways of seeing, understanding, interacting, and doing. Sustainability leaders are informed, aware, realistic, courageous, and personally hopeful in ways that genuinely attract others to the business of living collaboratively.
> *(Ferdig, 2007, p. 32)*

An emerging ecosystem of support for changemaking and sustainability leadership has emerged over the last decade in England. The following brief (and partial) survey traces the evolution of several strands of learning and leadership development. The connections between them emphasise the significance of trust, relationships, and openness in sustaining and nurturing sustainability leadership.

ROUTES TO RESILIENCE

Almost a decade ago, the Impact Trust, a north-south charity with deep roots in South Africa's transformation, led a significant effort to drive 'future-fit' education for sustainability leadership in practice, developing the Routes to Resilience programmes in a unique collaboration between the Impact Trust, the Cambridge Institute for Sustainability Leadership in South Africa (CISL-SA) and a wide network of contributions from leading sustainability experts around the world. In 2018, the Trust collaborated with United World Colleges and Ashoka Changemaker Schools to expand access to the Trust's Routes to Resilience programme materials in the UK. Some 24 educators from schools across the UK and Europe participated in a four-day residential immersion at Atlantic College to expand awareness of the sustainability lens and how to integrate it into educational and classroom practice. A cohort of youth from 21 countries engaged in a similar and parallel programme at Atlantic College, meeting in the middle to collectively share insights. For several schools, the sustainability lens led to innovation such as integrating the Sustainable Development Goals (SDGs) into curriculum planning (e.g. Barrowford, School 21, The Studio Liverpool). Routes to Resilience – Developing resilient leaders (https://routes2resilience.org) programme materials and training programmes are accredited for continuing professional development (CPD) and remain available in a 'gifting model' for supporting educators committed to future fitting education in sustainability.

NEW CAPABILITIES FOR A NEW WORLD

Sustainability leadership connects educators to themselves, their schools, and the wider systems they inhabit. This 'vertical' approach to professional learning that emerged from a pilot with 14 schools through the New Capabilities for a New World Programme supported by the Greater Manchester Combined Authority Skills Team, The Careers and Enterprise Company, Ashoka, and The Edge Foundation in 2022. This introduced 'ecoliteracy' as a new capability for school leaders and schools were introduced to Routes to Resilience materials. New Capabilities takes a developmental, or human-centred systems approach to school leadership development. Exploring change from a systems and human centred design perspective, school leaders started to integrate perspectives of self and schools, with their contribution to the wider system (McInerney and Kewalramani 2023).

ASHOKA'S TOGETHERNESS PRACTICE

Emerging from The New Capabilities Programme, The Togetherness Practice (no date) is an online leadership development framework and is free to access. The Togetherness Practice is designed to support transformative action in three domains: the growth of internal capacities; attending to external conditions and engaging in the relational field i.e. reflecting on how we collaborate. The research for this, based on interviews with social innovators and system leaders, suggested that changemaking is a collective endeavour. Relationships we forge help us grow the capacities we need to enact the contribution we want to make. The implication of this is that sustainability leaders, develop in communities, not in isolation. The Togetherness Practice is being used to support school and system improvement in several schools through the Institute of Education at the University of Worcester.

RETHINKING LEADERSHIP FOR SUSTAINABILITY

An ecosystem of practitioners and thinkers is emerging in the UK. The commonality is a more holistic, expansive view of leadership development to support a more integrated and holistic view of education. 'The Big Education Leadership Programme' offered by London-based Multi Academy Trust, Big Education (https://bigeducation.org/big-education-leadership-programme/)

supports school and system leaders to adopt more expansive view of education for all young people. ThoughtBox, (2014) have a leadership development programme that takes a regenerative approach. Portland Education (2017) have developed a successful 'Changemakers' sustainability leadership programme with The Eden Project. These innovations and the ecosystem driving them are creating conditions for sustainability leadership. However, these efforts will remain marginal as long as sustainability itself remains on the fringes of school development planning, professional learning, and leadership development in schools. With this bigger picture in mind, a collaborative network has emerged under the banner of 'Rethinking Leadership' (https://bigeducation.org/rethinking-leadership/). It seeks to encourage approaches to leadership development such as compassionate leadership and integrated leadership that are wider in scope, more grounded in individual purpose, and enable sustainable responses to the emerging needs of young people and the wider system.

Conclusion

Bregman (2017) points out that "The great milestones of civilisation always have the whiff of utopia about them at first." Over centuries, our agency and action in pursuit of 'progress' has distorted our relationship with nature; we have secured dominion over nature rather than live in balance with it.

At the heart of a sustainability leadership approach is our ability to question our relationship not only to nature but also to our worldviews of 'growth' and 'progress' and 'sustainable development' too. It is now imperative that we move away from a deeply held mindset of 'more for everyone' towards a more equitable worldview of 'enough for all' and sustainability leadership is a starting point for this transition. This is immediately uncomfortable because the status quo serves many of us well, yet it obscures the reality that we are depleting the natural capital on which our future standard of living is based. Over the coming decades it is likely that we will need to become comfortable with discomfort and discontinuity.

Our education system is the highest leverage change strategy we must make a difference in the time we have. Sustainability leadership, by each one of us, has the potential to be the fulcrum of this change. Sustainability leadership is about human potential and our will to harness it. Sustainability leadership is a capability that all education leaders need because it enables us to grow and shift our perspective from 'silo' to 'system.' The hope of this chapter is that our own praxis starts with us questioning some fundamental assumptions. Firstly, recognising that intelligence is socially constructed and socially mediated and thus not a private good, but a collective asset (Brown and Lauder, 2001). Secondly, accepting that meaningful 'progress' has not been driven by a neo-Darwinian, 'survival of the fittest' paradigm but by skilful collaboration that unlocks potential and contribution (Heffernan, 2014). And thirdly, rejecting the conventional wisdom that we can 'solve' problems without attending to the conditions that are holding them in place.

Hope does not spring eternal if the well is polluted. Maintaining a hopeful, positive approach may be an astute strategy, for hope may be essential for us to be able to grow the capacities we need that lie beyond our current capabilities. Embracing the discomfort of uncertainty can be a source of learning and growth (Coniff, 2022). A positive, hopeful mindset is important for our wellbeing too and to reframe our 'learned helplessness' into 'learned optimism' (Seligman, 2011).

Sustainability leadership invites a praxis of hope. It also points to something more profound: creating conditions for us to engage with an expansion of our consciousness that builds from the best of ourselves and one another as we search for new ways to forge meaning and purpose that lie beyond the material. Sustainability leadership is hopeful because it attends not just to our mind, but to our body, our heart, and our spirit. It embraces the cultural conditions we create when we act with minds of unconditional positive regard in relation to others, and it cultivates minds of love.

The hope, above all, is that we see schools create conditions that allow educators to enact a praxis of hope. In practical terms, encouraging educators to consider who they are becoming and the contribution they are making as they grow in competence 'to' do their roles. Education is fundamental to the process of human development and recognising this is the starting point for us to integrate what is good for the self, with what is right for children and young people in schools with what will support a shift in all our systems towards a world that can create 'enough, for all, for ever.'

References

Bregman, R. (2017) *Utopia for Realists*. New York: Little, Brown And Company.

Brown, P. and Lauder, H. (2001) *Collective Intelligence*. In Capitalism and Social Progress. London: Palgrave Macmillan, pp. 209–226. https://doi.org/10.1057/9780333985380_13.

Bryan, A. (2021) Pedagogy of the implicated: Advancing a social ecology of responsibility framework to promote deeper understanding of the climate crisis. *Pedagogy, Culture & Society*, 30(3), pp. 1–20. https://doi.org/10.1080/14681366.2021.1977979.

Chater, N. and George, L. (2023) The i-frame and the s-frame: How focusing on individual-level solutions has led behavioral public policy astray. *Behavioral and Brain Sciences*, 46, e147, pp. 1–84. https://doi.org/10.1017/S0140525X22002023.

Churchman, W. (1967) Wicked problems. *Management Science*, 14(4), pp. B141–B142, http://www.jstor.org/stable/2628678 .

Coniff, S. (2022) *Uncertainty experts – interactive learning documentary*. [online] www.uncertaintyexperts.com. Available at: https://www.uncertaintyexperts.com/ (Accessed: 17 February 2024).

Cook-Greuter, S. (2004) Industrial and commercial training making the case for a developmental perspective. *Emeral d Industrial and Commercial Training.* [online] 36(7). pp. 275-281. Available at: https://amielhandelsman.com/wp-content/uploads/2016/02/Cook-Greuter-Making-the-CaseDevelopPerspective-2004.pdf.

Cunliffe, A.L. (2016) Republication of 'on becoming a critically reflexive practitioner.' *Journal of Management Education*, 40(6), pp. 747–768. https://doi.org/10.1177/1052562916674465.

Department for Education (2023) *Sustainability and climate change: A strategy for the education and children's services systems.* [online] gov.uk. Available at: https://www.gov.uk/government/publications/sustainability-and-climate-change-strategy/sustainability-and-climate-change-a-strategy-for-the-education-and-childrens-services-systems.

Ferdig, M.A. (2007) Sustainability leadership: Co-creating a sustainable future. *Journal of Change Management*, 7(1), pp. 25–35. https://doi.org/10.1080/14697010701233809.

Freire, P. and Ramos, M.B. (1970) *Pedagogy of the Oppressed / Translated by Myra Bergman Ramos.* New York: Seabury Press.

Gilbride, N., James, C. and Carr, S. (2020) School principals at different stages of adult ego development: Their sense-making capabilities and how others experience them. *Educational Management Administration & Leadership*, 49(2), p. 174114322090372. https://doi.org/10.1177/1741143220903724.

Graves, C.W. (1974) Human nature prepares for a momentous leap. *Futurist.* April 1974, pp.72–87

Heffernan, M. (2014) *A Bigger Prize.* New York: Simon and Schuster.

Homer-Dixon, T.F. (2021) *Commanding Hope: The Power We Have to Renew a World in Peril.* Toronto: Vintage Canada.

Howard, P.G. (2018) Twenty-first century learning as a radical re-thinking of education in the service of life. *Education Sciences*, 8(4), p. 189. https://doi.org/10.3390/educsci8040189

Hy, L.X. and Loevinger, J. (2014) *Measuring Ego Development.* New York: Psychology Press.

International Commission on the Futures of Education (2021) Reimagining our futures together: A new social contract for education. New York: UNESCO. https://doi.org/10.54675/asrb4722.

IPCC (2023) *Synthesis report of the IPCC sixth assessment report (AR6) summary for policymakers.* [online] *IPCC.* Intergovernmental Panel on Climate Change. Available at: https://www.ipcc.ch/report/ar6/syr/downloads/report/IPCC_AR6_SYR_SPM.pdf.

Kahneman, D. (2011) *Thinking, Fast and Slow.* New York: Farrar, Straus, and Giroux.

Kegan, R. (1982) *The Evolving Self: Problem and Process in Human Development.* London; Harvard University Press.

Kegan, R. (1998) *In Over Our Heads.* London: Harvard University Press.

Laloux, F. (2016) *Reinventing Organizations: An Illustrated Invitation to Join the Conversation on Next-Stage Organizations.* Bruxelles: Nelson Parker.

Lawrence, M., Janzwood, S. and Homer-Dixon, T. (2022) *What is a global polycrisis? And how is it different from a systemic risk?* [online] Available at: https://cascadeinstitute.org/wp-content/uploads/2022/04/What-is-a-global-polycrisis-v2.pdf.

Lawson, A. and Read, R. (2022) *Activism, the very idea… Is the concept of activism itself a key block to mass action on climate?* [online]. Available at: https://systems-souls-society.com/wp-content/uploads/2022/04/Activism-the-very-idea....pdf (Accessed 1 February 2024).

Maslow, A.H. (1951) Higher needs and personality. *Dialectica*, 5(3–4), pp. 257–265. https://doi.org/10.1111/j.1746-8361.1951.tb01056.x.
McInerney, S., and Kewalramani, R. (2023) New Capabilities for a New World: leadership development for school and system improvement. London: The Edge Foundation.
Pollner, M. (1991) Left of ethnomethodology: The rise and decline of radical reflexivity. *American Sociological Review,* 56, pp. 370-380.
Portland education (2017) *Portlandeducation.* Available at: https://www.portland education.co.uk (Accessed: 15 November 2024).
Rees, W.D. (2023) Overshoot. *The Journal of Population and Sustainability*, 7(1), pp. 15–38. https://doi.org/10.3197/jps.63799953906865.
Routes to Resilience – Developing resilient leaders (2022). Available at: https://routes2 resilience.org/ (Accessed: 15 November 2024).
Rowson, J. (2017) *The unrecognised genius of Jean Piaget.* [online] Medium. Available at: https://jonathanrowson.medium.com/the-unrecognised-genius-of-jean-piaget-78c2914e306.
Sanford, C. (2020) *The Regenerative Life.* Boston and London: Nicholas Brealey.
Seligman, M.E.P. (2011) *Learned Optimism.* New York: Vintage.
Sterling, S. (2001) *Sustainable Education: Re-Visiting Learning and Change.* Devon: Green Books.
Thurston, A. (2022) Sacred design lab personal communication. Togetherness Practice (no date) *Togetherness Practice.* Available at: https://www.togethernesspractice.org.
Transforming Leadership Programme — ThoughtBox Education (2014) *ThoughtBox Education, ThoughtBox Education.* Available at: https://thoughtboxeducation.com/ leadership (Accessed: 15 November 2024).
Watkins, A. (2015) *4D Leadership.* London: Kogan Page Publishers.
Wilber, K. (2000) *A Theory of Everything: An Integral Vision for Business, Politics, Science, and Spirituality.* Boston: Shambhala.

11

LEADERSHIP BY ADAPTATION

A collaborative approach

Antoinette Frearson, Morgan Phillips, Emma Laurence and Sam Jackson

Introduction

By speaking with experts on sustainability in a conversation-captured style, this chapter picks up on the drivers of climate change discussed in Chapter 1 and looks at these drivers in the context of education. Firstly, the challenges and prevailing issues faced by the UK education sector are identified and opportunities for agency and action are explored. Later in this chapter Emma talks to Sam Jackson, Director of Climate Science and Impact at Ecologi, about the power of networks and meaningful collaborations within communities, and how this generates adaptive potential, but first Antoinette interviews Morgan Philips, from Global Action Plan on the specific challenges and opportunities for adaptations in schools.

Antoinette: Morgan, the examples in your book Great Adaptation (Phillips, 2021) that highlight how people can adapt positively to climate change are inspirational. As you see it, what challenges does climate change pose for the education sector in the intermediate and long-term?

Morgan: The Department for Education (DfE) and Ofsted are starting to recognise these challenges, and the need for acceleration and increased awareness to adapt, but there's still an underestimation of what lies ahead. The DfE has a remit for England and now has a sustainability and climate change unit. They have produced a strategy and non-statutory guidance (DfE, 2022) for education settings, but schools are being asked to achieve net-zero on little or no budget.

DOI: 10.4324/9781003506850-15

As we're seeing, climate change can increase the risk of extreme weather events such as heatwaves, drought, flooding, and forest fires. Sadly though, we're living in an age where some schoolchildren are still being told they can't wear shorts or take blazers off in heatwaves because it's not part of the uniform policy. Regardless of uniform policies, heatwaves are a threat and a safeguarding issue. If children or young people are outside in high temperatures, whether in shorts or trousers there's a chance they won't be feeling great anyway and if they haven't had the right amount of food or water, they may be prone to dehydration and fainting. So, one of the challenges schools are facing is the need to provide more shade and shelter in their outdoor spaces to protect children from extreme temperatures. Droughts, which can come along with heatwaves, can lead to water shortages, and this can – when extreme – impact on the availability of food and water. There's also potential, even here in the UK, that droughts and heatwaves lead to insect infestations which we haven't seen yet but occur in other parts of the world. Flooding can also be a safeguarding issue as sometimes floodwater rises slowly, but other times it comes in a deluge which could flow through a playground or into school buildings, and could interrupt the water supply or wastewater removal, causing the school to close. Another risk from an increasingly hotter and drier climate is forest fires. If even just a small woodland catches fire which then spreads to a school, it could have devastating consequences and it's not just the disruption and physical impact on the school but also the fear and trauma – the emotional impact – for staff and families. These are all safeguarding issues in the intermediate and longer-term, and schools need to look at the potential impacts of extreme weather on their locality and how they can build in some resilience.

Another challenge facing young people is climate anxiety (Pihkala, 2020). They are, understandably, becoming very worried about their futures, the people they care about as well as humanity and endangered species around the world. As we proceed further into climate breakdown, this anxiety will likely grow [see Chapter 2]. It will become a challenge for schools; they'll need to be able to manage that and help young people cope with it, and to have a positive response by giving them agency. This means that educators, are becoming increasingly aware of the need to become climate literate themselves; to understand the causes and threats; what sustainability is, and how to deliver on adaptation, resilience, and sustainability education. Already busy teachers are having to grasp all this, as it's likely not in their initial training and much sustainability education is not pedagogically the same as what they're doing now. Schools need to have access to training and resources so staff can feel more confident in these areas.

It's important that schools recognise the difference between incremental and transformative approaches to adaptation too. Incremental adaptation involves steady and small adjustments to the impacts that are happening, but this can be problematic because it often masks and underestimates the scale of the problem, rendering the adaptation plan redundant sooner than expected. In that sense, incremental adaptation can be a form of maladaptation which leaves future generations more acutely exposed to climate impacts when the adaptation plan fails them. Transformative adaptation situates itself within a wider, more fundamental effort to tackle the root causes of climate change (Read and Phillips, 2024). That's the difference, and an important consideration when you're thinking about developing an adaptation strategy – incremental adaptation will only take us so far, and given the trajectory we're on with climate change, it might be that a more transformative outlook is now essential. There is inspiring work going on through the National Education Nature Park initiative (Natural History Museum, 2025, the Climate Action Awards (DfE, 2023), Sustainability Support for Education (DfE, 2024) and the Climate Ambassadors programme (STEM Learning, 2022; University of Reading, 2024). More investment is needed for these projects but it's a start, and a base to grow from.

Antoinette: Tell me about the need for agency in schools.

Morgan: It's so important that young people who are graduating or leaving school are prepared to take collective action for the good of people and the planet. Currently, the English education system is designed to preserve and enhance the status quo, rather than transform it. Consequently, certain world views and values are being activated then reinforced in our schools. This translates to a set of outcomes in terms of the knowledge, skills, and motivations that young people graduate from school with. Professor Philip Woods argues that education has emphasised individualism, the "promotion of an 'enterprising self' and the relentless measurement of people against targets and performance criteria" (Woods, 2012). The education system, therefore, cannot be thought of as benign in terms of its impact on the climate crisis, it is currently nurturing generation 'consumer' when what the world needs right now is generation 'action.' To expand on this a little bit, we're entering a period of climate uncertainty where we don't know how much the world is going to warm by. It could be anything between 1.5 degrees and 4 degrees Celsius this century. Maybe higher. One approach is to prepare young people to be adaptive in a reactive way, with the goal being to survive a climate breakdown scenario in an individualistic way. Alternatively, we nurture young people's willingness and ability to create change, where the goal is to nurture their sense of agency and their desire to act collectively in socially and environmentally just ways.

Maybe what we need is education that isn't *for* anything at all. In other words, we need more of an emphasis on education for education's sake; a process of self-discovery of ourselves and the world, to learn in ways that are more self-directed and more engaged which will lead to better outcomes in terms of what you do when you leave school. Students need more control over what they're learning and how they're learning. They need more agency, so that they are better equipped to adapt to climate change in a transformative, rather than incremental way. I'm not saying students should be completely left to their own devices; they still need guidance from teachers, and teachers themselves need guidance and training too. If we can step toward more local and hyper-local control over what is learnt and how it's learnt, it will give some of that control back to the teachers and to the children themselves so they can shape their own learning. I think we can trust young people and teachers to do this, most people, deep down, are decent – few would squander such freedoms. It's worth remembering that, currently, we are forced to trust a select few people in Westminster, and they're often held in check by powerful vested interests. So, if what we want is a school system that nurtures a generation of young people who can face uncertainty with a sense of agency and a transformative adaptation mindset, we need to think more holistically about how the education system is run and what its purposes are. I would like to see it become less instrumentalised, and less about economic growth, or even 'sustainable development' for that matter. If questions of purpose were more devolved, we would have an education system that is a lot more dynamic and more diverse in terms of the content, pedagogy, and outcomes. We wouldn't then have a generation of young people who all have similar beliefs, and all know the same things. We would have a generation of people who have, between them, a massive array of knowledge and skills. That would surely put us in a better position to cope with climate uncertainty and more resilient as a society.

Antoinette: Tell me about the opportunities for agency in schools.

Morgan: The new DfE support hub (2023a) promises to be helpful in that it will enable schools to do a self-assessment around scope 1, 2, and 3; carbon emissions, climate education and other things, including adaptation and resilience. This will ask them questions around the geography of the school, the quality of their buildings, their playground, and other outdoor spaces, and this will allow them to receive an assessment score on making adaptive changes. They'll then be signposted to resources and training opportunities from companies, local authorities, and non-governmental organisations (NGOs), who will help support schools to create an adaptation plan as part of their wider Climate Action Plan.

We know that many UK schools today are in a poor state of repair, and we must hope that a 'build back better' mantra is adopted by them and the government as they make major investments in the coming years. It's an opportunity not to be missed in both climate mitigation and adaptation terms. Schools need to work with climate literate architects to ensure that things like the direction windows are facing in a school are thought about for heating and cooling, and how school buildings can be designed to ensure they're allowing for natural ventilation – to lessen the need for air conditioning on hot days. It is obviously a chance to install solar panels and retrofit heating systems to make them greener. But it's also a chance to improve the drainage systems, erect flood walls, and plant trees – for shade and air quality. This is a learning opportunity for students too. Children can get involved in measuring the room temperature with the windows shut compared with the windows open, so they can develop their own behaviour changes. Other strategies could be to install shutters to block out direct sunlight in the summer months, or to replace concrete with grass or vegetable plots.

When it comes to adaptation, the strategies need to be quite bespoke as schools are all situated differently. To start with, they need an assessment of the risks that they are likely to encounter which will involve research into the data that's available such as local sea-level rises or weather patterns – again this is something that could be student led. By doing this, schools can begin to understand how many days they are likely to experience heatwaves, storms, droughts, flooding or even forest fires. They can then decide what adaptation strategies they'll need to put in place, which won't all be about making physical changes to the school. It might be a decision to reschedule the school day so that children aren't involved in tasks requiring intense concentration, or physical effort at the hottest times. At a national level, the DfE may wish to consider moving high-stakes exams to the cooler months of the year – revising for and performing in exams is hard enough anyway, doing it during a heatwave makes it doubly difficult. Opportunities for other transformative changes include looking at heavily flood-prone sites or those with sea-level rises to decide if the whole school should be rebuilt or merge with another school elsewhere. Strategies such as these could inform school leaders to help lobby for and implement changes.

Antoinette: How can we empower students to tackle these deeper changes by using their voice in school and in their local community?

Morgan: It's important that young people aren't having adaptation and mitigation strategies inflicted upon them. It's important they have a say and are involved in planning the adaptation or mitigation strategy and helping to implement it. It's an unmissable learning opportunity; we're talking project management, research, communications, budgeting, creativity, all these skills and more can be developed. More practically, engaging students is important for buy-in too. A lot of work around adaptation and mitigation is still going to be dependent on people's behaviours so unless you're involving the young people – whose behaviours will need to change in the process of designing that change – then they're not going to embrace the changes as quickly. Schools must recognise the need to involve their students, not just as a one-off learning opportunity but to ensure they are understanding the need for the changes that are happening around them and having a genuine voice in those changes. Additionally, through learning what's happening in a school and witnessing it for the first time they're well placed to help that parent who runs a shop, hairsalon, garage or accountancy firm. Anything, that can help them with those mitigation and adaptation opportunities.

Antoinette: Can you tell me about any schools who are pioneering the approaches that you've mentioned?

Morgan: I know there are schools engaging young people in project-based learning by getting them involved in real-world projects through programmes like Good Life Schools (Global Action Plan, 2024). It's exciting when you see the young people being given a project, learning through doing it, being allowed to make mistakes and not being afraid of things failing; it's the creativity and imagination, the critical thinking, and the teamwork which is building them up to be 'Generation Action,' the kind of people who want to come into the world to create changes after they've had a chance to develop and practice those skills. One of the big barriers they're overcoming is the feeling of being alone. They're seeing the other people actively care about these issues too and recognising the power of collective action. Moreover, I think teachers talking about climate issues, visibly acting on them and enabling others to act, shows young people – especially those suffering from climate-anxiety – that they're not alone in those fears. It gives them confidence to start trying to act themselves, rather than being worried that they're the only one who cares and that their efforts are going to get lost. So, that's where I'm seeing the potential, in giving people the freedom, the confidence and the permission to act, especially in groups. They can then develop into the change-makers we need. It would be lovely to see more examples of adaptation. I hope there'll be more coming.

Antoinette: Yes. I'd like to see a national forum set up for schools to talk about these climate change challenges and opportunities, enabling an inspirational knowledge exchange for embracing adaptation; a place to reflect on the deeper changes needed in schools, the education system and the wider society – ensuring students (our future citizens) are involved at every stage too. I hope this conversation has inspired educators to start their own conversations with students, colleagues, and leadership teams, and together, take action to become those change-makers that the world desperately needs today and tomorrow.

The need for collaborative action in schools to respond to the climate crisis

While the impact of the turbulent context we find ourselves in is not to be underestimated, out of crises often come stories of hope and opportunities to learn about how we might nurture our adaptive capacities and create powerful communities in the face of uncertainty. Schools often serve as the heart of their communities and the education sector has long been one characterised by a context of "rapid and relentless educational change" (Steward, 2014, p. 52). This highlights how well positioned educators and educational leaders are in supporting the adaptive capacity of their communities. Leadership literature has already identified adaptive leadership as the approach for the knowledge era (Uhl-Bien et al., 2007) but within western cultures individual perspectives are prioritised over the collective. Despite this however, our success as a species is so often attributed to how inherently social we are. We have all seen those goosebump inducing videos where strangers come together to lift a train carriage off someone who is stuck in the gap. When you look closely, most stories of hope involve people relying on each other. Be it the quiet support of one great relationship or an inspirational mass movement, people need people. Therefore, we should consider how relationships might influence any action we take in response to this book. To do this, we foreground a perspective on leadership as an interaction *between* people rather than a quality which resides within a few.

Research findings show that social connections help people to regulate their emotions, cope with stress and remain resilient in the face of adversity

(Holmes et al. 2020; Van Bavel et al., 2020). Szreter and Woolcock (2004) identify a three-dimensional framework outlining the different kinds of social capital which support us in understanding aspects of our social environments. These include:

- bonding social capital: trusting and co-operative relations between "people like us" who are "in it together" (Claridge, 2018, p. 2)
- bridging social capital "between people who know that they are not alike," (Szreter and Woolcock, 2004, p. 655),
- and linking social capital, which they define as the "norms of respect and networks of trusting relationships between people who are interacting across explicit, formal or institutionalized power or authority gradients in society" (Szreter and Woolcock, 2004, p. 655).

Putnam, (2001) suggests that it is not simply the numbers of social connections which provide social capital, but it is the norms of reciprocity and trust within them that do so. It is one thing to know someone, but to want to support them and to know that they would seek to do the same turns an association into a vital source of resilience and strength in times of challenge. The reciprocal nature of these relationships mean that both parties benefit from this ability to draw on each other as a soundboard, a confidante, a broader well of expertise and experience and as a conduit for resources. Developing leadership from a network perspective then, is about promoting the social capital within systems: promoting cooperation, mutual respect and trust between people in order to promote whole community efficacy. Next, Emma discusses with Sam Jackson, Director of Impact at Ecologi, how these networks impact on his work in regeneration projects across the globe.

Emma: Sam, what relevance do you think interpersonal and inter-community relationships have, when it comes to implementing climate solutions?

Sam: They are incredibly relevant, because they form the larger part of the solution to the climate crisis. There are a lot of ways for people to interpret potential solutions to climate change very differently, and a lot of things we could disagree about (Hulme, 2013). That's not to mention the complexity of the challenge itself, which can also be understood and scrutinised in countless ways. It can feel excruciating to appreciate the relatively straightforward physical solutions to this challenge (reductively: to stop producing greenhouse gas emissions; to conserve and restore nature; and to deliver a just transition to a less- or non-extractive global economy) while bearing witness to a staggering societal failure to deliver on those solutions. Part of the problem is that these solutions are,

> naturally, a lot easier to talk about than to implement. But I think the other component is that often our approaches to solving the problem are rooted in irrational mindsets, and we consciously or unconsciously 'adapt' by making comfortable but erroneous assumptions about the future. One example of this is when people subscribe to the belief that just around the corner, we can expect to conjure up some kind of one-shot, top-down fix to climate change. Usually, this misguided mythology is perpetuated by the systems which make us feel that we can build our way out of a crisis using the same means that caused it in the first place. It usually appears as an expectation of some kind of future technological intervention – such as the hyper scaling of tech-based carbon dioxide removal, or the deployment of potentially-dangerous geoengineering, or perhaps some unspecified miracle of artificial intelligence (AI).

Each of these technologies will play some role, there is no doubt. But thinking of them as singular 'techno-fix' solutions for solving the climate crisis is wrong, because the climate crisis is not a singular problem. It is not a problem that can be solved with neatness or finality. Instead, it's a fiercely complicated phenomenon which we can work on in patches, and make progress through effort and time and care. Adopting the technocratic mindset makes us believe a false narrative that we'd solve everything, if only we could build the technology, and it delivers us directly to either of two easy-but-incorrect conclusions: that everything is fine, because we can build the sacred technology; or that we are all doomed, because we can't. Neither these conclusions stack up when we actually appreciate the gravity, the complexity, or the source of the challenge we are trying to address.

Because of the nature of the problem, impactful solutions to the current crisis will be discovered bottom-up, not top-down. This means we need to look inward to start to understand how the problems of climate change and environmental breakdown might be addressed. That's inward into ourselves, but also into our communities, our social groups, our workplaces, the societies to which we belong – and of course, the schools which educate our children, who will be left to cope with climate change for long after we ourselves are gone. In this way we can examine and explore climate solutions locally – and organically grow this local action everywhere. I've heard this kind of approach called 'glocalisation' (as contrasted with 'globalisation' for the top-down approach). And while there is no miracle technology coming to save us, the levers of change are in our hands regardless. We caused the problem of climate change through building extractive and exploitative political and economic structures, which we often consider as if they are immutable – as if they govern the laws of the universe itself. But they aren't, and they don't. At the same time, we often talk about the climate crisis as if the only thing that matters is reducing carbon emissions, and this isn't true either. We have concurrent crises of climate change, biodiversity collapse, over-consumption and many others – all interconnected, and all driven by the momentum of the globally extractive systems we have built. It follows that implementing a global change on this kind of scale will only come about through a truly collective action, a sort of coming together to achieve communal ends – which I believe, when we zoom in a bit, is weaved out of the fabric of interpersonal and inter-community relationships.

Emma: How do trust, collaboration and reciprocity fit into the work of restoring nature at Ecologi?

Sam: Trust, reciprocity, honesty, accountability, transparency – these are all foundational concepts for our work at Ecologi, which also represent (some of) the necessary conditions for its success. Without collaborative, constructive, and sometimes deeply personal bonds of partnership – which necessarily transverse continents, language barriers, and generational and cultural differences – there is no scaleable route to the conservation and restoration of nature. This essential principle of constructive partnership applies at both micro- and macro-scales: from the intricate logistics and coordination of people during project implementation, all the way up to the organisation and delivery of the large-scale deployment and transfer of funds from international investors and donors.

In my work, I am fortunate to occasionally visit these restorative projects with my team, for the purposes of partnership-building and project due diligence, and to bring back deeper understandings of the successes, challenges and realities of the projects. One such visit was to Vilcanota, in Perú, with a wonderful initiative called Acción Andina. This initiative, and the social movement it represents, achieved deserved recognition in being awarded the Earthshot Prize in 2023, and being named as one of seven UN World Restoration Flagships in 2024 (see case study below).

Most of the people who run or work for purpose-led organisations do so in good faith, and because they care deeply about the cause, whatever it may be – so much so, that they have dedicated their labour and their company's resources to contributing to it. I've enjoyed countless illuminating conversations with people who might be considered 'competitors' – conversations made up of little things, like sharing stories, reflecting on the challenges we've each faced, and the sensitive asking of questions. I don't have all the answers to everything, and neither do they – but if we talk, and share, and laugh together, it's likely we can learn something. Perhaps we'll uncover something together which neither of us would have come up with on our own. The neoliberal narrative that we ought all to be suspicious of one another is, generally, holding us all back. This recognition of the value in collaboration is the exact same thrust which leads environmental scientists to build extensive (and extensively valuable) free-to-access, open databases for biodiversity data, climate model output, and research articles. Doing so allows others to use our resources to help us all collectively work on the problem. Anyone who's ever attended a conflict resolution seminar may recognise another microcosmic example of this: reframing the conflict situation so as to resist the temptation to think about you versus me and instead consider it as you and me, versus the problem.

In all of these cases, we are observing an abstraction from the singular means to the communal means – which is logical, since we are pursuing a communal end. Solving the climate crisis and restoring balance to our relationship with nature is the ultimate communal end, so why shut ourselves off from pursuing it with communal means? In my view, understanding and internalising the Andean concepts of mink'a and ayni [see case study below] could help us learn something important: that trust, collaboration, and reciprocity are essential, and that coming together in our communities is effective in achieving our collective aims.

Emma: How does global-scale climate action rely on social capital, and interpersonal relationships?

Sam: Well, the chief vehicle for delivering legislative and climate policy change internationally is the UN Framework Convention on Climate Change (UNFCCC) and its Conferences of the Parties (COPs). These are annual fora where world leaders and negotiators come together to agree and announce environmental pledges and policy initiatives. At the end, a 'cover decision' is published: a UN-level text which represents the consensus-based outcome of the COP, and often the host country will adorn it with a fancy name like the Glasgow Climate Pact (from COP 26), or the Paris Agreement (from COP 21). These cover texts provide a summary of the year's international political sentiment on climate change, and act as a lag indicator for the more rapid pace of change which is underway elsewhere in society. While it is rare for the Parties to legally bind themselves with these agreements, they still carry a tremendous weight, acting as both a bellwether for international political will, and as a powerful influencer of economic heft.

Much of the pledge-making and policy design, as well as many of the peripheral coalitions which are announced during the COPs, are carefully and systematically crafted into shape beforehand, and during the conference in UN negotiating rooms, by seasoned negotiators. But, especially in recent years, when the negotiations overrun, the final wording of the cover decisions are just as often negotiated in the small hours of the morning by frantic, private 'huddles' of politicians and ministers, dashing between meeting rooms. In these contexts, the COP outcomes can be prescribed by the interests of the more vocal or economically-dominant Parties – Parties represented by individuals who find themselves in the right positions, in the right rooms, with the right people, and with the loudest voices.

In terms of the different kinds of social capital and how they can be leveraged to deliver progress, the climate crisis allows at least two ways to think about this. The first is that it shows us how we are different, which provides an opportunity to bridge the gaps between us. We are not all equally responsible for causing climate change, and neither are we equally going to feel its impacts (IPCC, 2022). Also, as national and subnational communities, we don't all have the same economic means, geographies, or all the same social norms. Politically, this understanding manifests in the core UNFCCC principle of common but differentiated responsibilities ('CBDR'), where all countries have a shared responsibility to act on climate change, but those responsibilities are not all the same: due to these disparities in wealth, vulnerability, historic responsibility, and so on. This principle ties us all together, but at arm's length: simultaneously we are (1) all the same, since we have a shared goal and a shared responsibility for reaching it; and (2) all different, since we have different capabilities, responsibilities and obligations. This is one way in which we form the weaker bond of 'bridging' social capital.

The stronger bond is found in the way the climate crisis shows us how we are all the same. There are plenty of shared narratives of identity which we can leverage collectively – whether that be in the negotiating halls of the UN Conferences, or in our homes and in our schools. We won't all be affected by climate change equally, but we will all be affected, and there are many values and norms we share which transcend cultures, languages, generations and political barriers. Earlier I mentioned trust, reciprocity, honesty, accountability and transparency. I'd now add others: dignity, respect, compassion, fairness, freedom, peace. These are widely-held values. They are certainly values held by enough people on the Earth who, in coming together, could move the dial on climate change. And regardless of what we believe, we can still find narratives in which we share the most fundamental characteristics of our nature with all other people of the world. The destructive behaviour which we have inflicted upon the Earth is to all of our detriment (and that of all life). This Mother Earth is our only home; she is the only home that has been known to any human in history – and so far, as we have thus far proven, any life, anywhere. And in our neglectful and extractive behaviour over recent centuries, we have been slowly but assuredly poisoning her.

Whether or not we in the West currently acknowledge the more traditional narratives about our relationship with the Earth, whatever factor *is* the most important in one's life, it will be affected by climate change. This applies, whether that thing you care about is money, or family – or more specific, oft-politicised topics, like housing, education, religion or immigration. It doesn't matter. The climate and other related crises have enormous implications for all of it. So whichever of these is our most important priority on a personal or spiritual level, we all have a stake in the climate crisis – and that means there is opportunity to find common ground between us, and build crucial social bonds upon it. In our hurried pursuit to conserve and restore what remains, with no miracle techno-fix just around the corner, returning to these broader and deeper shared narratives will help us to orient, to learn about where we come from, and where we must go now. They will prove to be the solid foundation for making meaningful progress to restore and renew our relationship with the Earth. This applies to whoever we are and whatever kinds of room we find ourselves in – from school classrooms, to UN negotiation rooms.

Case study: Vilcanota, Perú

During our visit, we were invited to participate in a Queuña Raymi – a festival of tree-planting, where hundreds of Indigenous people from rural communities in the High Andes come together to plant thousands of native Queuña trees (species of the genus Polylepis, though other native species are also planted during a Queuña Raymi) in a single day. The mountaintop glaciers which have provided communities with water for centuries are melting at a terrifying pace (see Dussaillant et al., 2019) and so healthy Polylepis ecosystems high up in the Andes are vital in helping to retain rainwater and keeping rivers and streams flowing. For local people, the motivation for planting these trees to restore these ecosystems is not a transactional matter of being paid to do so by others, it is a matter of doing what can be done to safeguard the future for the coming generations. Appropriately, Acción Andina focuses not just on planting trees, but also provides investment in leadership and capacity-building, community and social development, and conservation of existing ecosystems. The aim, is to capacitate a multigenerational effort of restoration and of communities coming together, right across the Andes, which will persist and thrive long after we ourselves are gone.

Queuña Raymi is also a celebration of Andean culture and community. The local people who take part are Quechua-speaking descendants of the Incas, and the Acción Andina team highlighted to us that Queuña Raymi exhibits two traditional Inca forms of work, mink'a and ayni (see Mannheim, 1986). Mink'a is community work: the carrying out of often voluntary work with communal means, and communal ends. It might be manifested as collective labour for building municipal infrastructure, for example. Ayni is reciprocity, a notion of mutual exchange of effort and energy: if I support you today, you might support me tomorrow. It has roots in more fundamental conceptualisations of the reciprocal nature between nature and humanity (Mannheim, 1986). These powerful intra- and inter-community narratives are often deeply neglected in the Western world, but I believe they can play an understated yet profound role in our global attempt to mend our broken relationships with each other and with the Earth.

Conclusion

The escalating frequency of intense weather events are having a significant impact on our everyday lives. As discussed with Dr Morgan Phillips, extreme heat can be a safeguarding issue for the young, old and vulnerable; drought can threaten the availability of food and water for all, and flooding events can be sudden and destructive. There is an urgent need to adapt to our changing climate which is inescapable and to do that successfully and effectively, it is imperative to change our thinking at every level.

Schools are important places to nurture change. Collaborative working with students and staff can increase the school's resilience to extreme weather events in myriad different ways. For example, making uniform policies more reactive to extremes of heat, creating areas of shade and shelter to keep pupils cool and dry, and developing risk assessments which assess the impact of these events on the school's specific locality. Additionally, comprehensive training and resources will assist staff in becoming more effective at supporting students with rising eco-anxiety and empower them to recognise the opportunity for agency and action, enabling them to become change-makers and building upon the pioneering work of other pioneering initiatives. It is hoped that students will be inspired to make links with their local communities and work together to champion greener lifestyles.

This sense of collaboration and community was further discussed with Sam Jackson from Ecologi, who identified the benefits of creating a mutual exchange of effort and energy. This can not only harness the power of networks by making meaningful collaborations and relationships, but it also enables the generation of adaptive capacity, which can be transformative on a local and global scale. We cannot underestimate the values of trust, collaboration and reciprocity in our important work on tackling the main drivers of climate change and its potential consequences.

A heightened awareness of 'how we are different' gives us an opportunity to recognise the disparities in wealth, geography, development and vulnerability, not only globally but in our local areas, which shows us that we will not all experience climate change equally. Nevertheless, our co-habitation of this planet and our shared experience of Mother-Nature, demonstrates 'how we are the same,' and it is this dual awareness of our situation which must strengthen our determination and motivation to act collectively, for the protection of all now, and for generations to come. Therefore, we must become 'Generation Action,' because in the face of climate change, we truly are stronger together.

References

Department for Education (DfE) (2022) *Sustainability and climate change: A strategy for the education and children's services systems*. Available at: https://www.gov.uk/government/publications/sustainability-and-climate-change-strategy/sustainability-and-climate-change-a-strategy-for-the-education-and-childrens-services-systems (Accessed 30 October 2023).

Department for Education (DfE) (2023a) *The support hub*. Available at: https://educationhub.blog.gov.uk/ (Accessed 30 October 2023).

Department for Education (2023b) *About the climate action awards*. Available at: https://www.educationnaturepark.org.uk/climate-action-awards (Accessed 16 January 2024).

Dussaillant, I., Berthier, E., Brun, F., Masiokas, M., Hugonnet, R., Favier, V., Rabatel, A., Pitte, P. and Ruiz, L. (2019) Two decades of glacier mass loss along the Andes. *Nature Geoscience*, 12, 802–808. https://doi.org/10.1038/s41561-019-0432-5

Global Action Plan (2024) *Generation action*. Available at: https://www.globalactionplan.org.uk/our-work/generation-action (Accessed 16 January 2024).

Holmes, E., O'Connor, R., Perry, V.H., Tracey, I., Wessely, S. and Bullmore, E. (2020) Multidisciplinary research priorities for the COVID-19 pandemic: a call for action for mental health science. *The Lancet Psychiatry*, 7(6), pp. 547–560.

Hulme, M. (2013) *Why we disagree about climate change*. Cambridge: Cambridge University Press.

IPCC (2022) Climate Change 2022: Impacts, Adaptation and Vulnerability. Available at: https://www.ipcc.ch/report/ar6/wg2/downloads/report/IPCC_AR6_WGII_FullReport.pdf. (Accessed on 24.02.25)

Mannheim, B. (1986). The language of reciprocity in Southern Peruvian Quechua. *Anthropological Linguistics*, 28(3), pp. 267–273. https://www.jstor.org/stable/30027956

Natural History Museum (2025) Natural Education Nature Park. Available at: https://www.nhm.ac.uk/about-us/national-impact/national-education-nature-park.html (Accessed 23 February 2025)

Newman, R. and Noy, I. (2023) The global costs of extreme weather that are attributable to climate change. *Nature Communications*, 14, p. 6103. https://doi.org/10.1038/s41467-023-41888-1

Phillips, M.(2021) *Great Adaptations*. Bristol: Arkbound

Pihkala, P. (2020) Anxiety and the ecological crisis: An analysis of eco-anxiety and climate anxiety. *Sustainability*, 12, p. 7836. https://doi.org/10.3390/su12197836.

Putnam, R.D. (2001) *Bowling alone: The collapse and revival of American community*. New York: Simon and Schuster.

Read, R. and Phillips, M. (2024. *Transformative Adaptations*. Portsmouth: Permanent Publications

STEM Learning (2022) *Bringing climate experts into schools*. Available at: https://www.stem.org.uk/news-and-views/news/climate-ambassadors#:~:text=Through%20the%20STEM%20Learning%20website%20and%20local%20networks,STEM%20Ambassador%20programme%20is%20funded%20by%20the%20UKRI. (Accessed 23 February 2025)

Steward, J. (2014) Sustaining emotional resilience for school leadership, *School Leadership & Management*, 34(1), pp. 52–68. https://doi.org/10.1080/13632434.2013.849686

Szreter, S. and Woolcock, M. (2004) Health by association? Social capital, social theory, and the political economy of public health. *International Journal of Epidemioliogy*, 33, 650–667. doi:10.1093/ije/dyh013U.

Uhl-Bien, M., Marion, R., and McKelvey, B. (2007). Complexity leadership theory: Shifting leadership from the industrial age to the knowledge era. *The Leadership Quarterly*, 18(4), pp. 298–318. https://doi.org/10.1016/j.leaqua.2007.04.002

University of Reading (2024) *University strategy: Engagement*. Available at: https://www.reading.ac.uk/about/strategy/engagement (Accessed 16 January 2024).

Van Bavel, J., Baicker, K., Boggio, P., Capraro, V., Cichoka, A. and Willer, R. (2020) Using social and behavioural science to support COVID-19 pandemic response. *Nature Human Behaviour*, 4(5), pp. 460–471.

Woods, P. (2012) Out of the Dragon's Den: Self-organising change for a broader vision. Available at: https://policypress.wordpress.com/tag/instrumentalising-trends/ (Accessed 16 January 2024).

12

HOPE FOR THE FUTURE

Communities, children and young people as leaders

Janet Harvell and Vicky Whitworth

Introduction

A well-known Native American saying states "We do not inherit the earth from our ancestors; [but rather] we borrow it from our children" (Gilbert, 2015). Talking with children and young people (CaYP) it is clear that they are anxious about climate change, which is unsurprising when we understand that they are most vulnerable to the effect of climate change, as explored in Chapter 2. This is not just about experiencing its impact, but also what this means for their everyday lives, and the changes that will need to be made in order to limit these effects. Research has demonstrated that this anxiety is being reflected in depression, sleep problems, distress, grief, anger, and feelings of helplessness (Clayton et al., 2017, cited in Sanson, Van Hoorn and Burke, 2019). A survey of British 11-to 16-year-olds identified that 74% were worried about the impact of climate change on their future (UNICEF UK, 2013). It is unmistakable that young people have a vested interest in "getting it right," fighting for social justice while holding the government to account, and the power of these combined voices has been hard to ignore. This is validated by Hickman et al. (2021, p. 1) who write that climate anxiety, and feelings of betrayal were linked to "perceived inadequate government response."

This would be an appropriate point to refer to the Swedish activist Greta Thunberg, an iconic example of empowered youth who, as a 15-year-old schoolgirl, sought to bring attention to the climate crisis and influence the upcoming Swedish election. Taking her protest to the heart of government, she committed to engaging in weekly school strikes every Friday until the government fulfilled its part of the 2015 Paris climate agreement (Watts, 2019). Holding governments accountable for the limited progress being made on

DOI: 10.4324/9781003506850-16

climate change, she galvanised young people across the world, providing an element of hope, where they may have previously felt powerless. An articulate and fervent communicator, she continues to speak forcefully about how the world her generation will inherit will be disproportionately harmed by the inaction of world leaders; holding governments responsible for the state of the world being left to young people to inhabit.

She inspires young people across the world to take positive action, such as lobbying governments for policy change, and such activism has proven to be an effective means in gaining agency. As Greta Thunberg stated,

> We can no longer let the people in power decide what is politically possible. We can no longer let the people in power decide what hope is. Hope is not passive. Hope is not blah, blah, blah. Hope is telling the truth. Hope is taking action. And hope always comes from the people.
>
> *(Youth4Climate Summit, Milan, Italy, 28 September 2021)*

We believe there is hope. Previous chapters have examined a range of responses to the environmental challenges being faced by the world. This chapter showcases some of the initiatives that are being implemented at national and international levels either led by CaYP, or having a positive impact on their lives, providing a beacon of hope. What is not in doubt is that the world is facing a serious existential crisis, and in many cases the political leadership remains wanting. A clear ethical binary has emerged in society, with fossil fuels and intensive agricultural practices being the villains and environmentalism the hero, yet oil production continues to grow. It is too simple to hold industrial nations such as China and India to account when considering the production of greenhouse gases, while wealthier nations such as the UK and US sustain this demand through consumerism. As Thunberg's rallying cry reminds her young followers, "…hope always comes from the people" (Thunberg, 2022).

At the same time there is a need to support poorer countries to prepare for the challenges of climate change, many of whom have rapidly increasing populations. With three-quarters of the world's population living in Asia and Africa, it is the CaYP from these nations who will be the ones to bear the brunt of disasters linked to climate change. The Loss and Damage fund for developing and vulnerable countries which was agreed on at the 2022 UN Climate Change Conference (UNDP, 2024) in Egypt will require realistic, internationally agreed responses and resources, not least financial support, for those countries which are particularly vulnerable to adverse impacts of climate change.

Global initiatives – child-centred approaches to climate action

The UN is at the forefront of galvanising member countries to commit to action and in 2015 member states adopted the 2030 Agenda for Sustainable

Development, identifying 17 Sustainable Development Goals (SDGs) (UN, 2023). In contrast to previous initiatives which worked to provide education for all and the eradication of poverty, the SDGs are more ambitious and included commitments towards Responsible Consumption and Reduction (SDG 12) and Climate Action (SDG 13).

Many CaYP are taking an active role in implementing strategies to address some of these issues. In their 2015 document 'Children and the Changing Climate,' UNICEF (2015) cites examples of how it has worked with CaYP from a number of countries being affected by climate change, developing bespoke programmes to provide the skills and resources needed to address some of the key challenges facing individual communities. These include the drilling of wells, how to stay safe during climate related disasters, climate-proofing schools, involving CaYP in decision making and becoming climate ambassadors.

In the Central African Republic former child soldiers have been retrained to manually drill for wells within their local districts. As well as adapting to climate change by sourcing fresh water, it has also been very successful in rehabilitating street children and demobilised child solders back into the community and providing them with valuable life skills (UNICEF, 2015, p. 7). In Uganda there are similar challenges related to changing weather systems and the impact on water supplies, with 36% of children still having to walk more than one hour each day to collect water (UNICEF, 2015, p. 23). Collaboration between the government and private sector has resulted in the building of solar powered pumps to bring clean water to the communities and enhancing the lives of CaYP.

In Honduras attention has been focused on educating schoolchildren and their families to support them in monitoring potential disaster trouble spots, and then identifying safe escape routes to be followed in the event of extreme weather occurrences (UNICEF, 2015, p. 13). There is a similar programme in Madagascar to enable children to be better prepared for weather extremes, which includes cyclone-proofing schools and developing "disaster resilient water systems" (UNICEF, 2015, p. 17). This is particularly relevant in Madagascar which has been identified as "the fourth most climate change affected country worldwide" (UN, 2024). Deforestation has played a significant part in the challenges now facing Madagascar, and to address this the UNICEF programme has focused on the use of locally available sustainable materials. Finally, Vietnam and Zambia have involved CaYP directly through focus groups and discussions. In Vietnam this has included listening to CAYP and documenting the impact that climate change has on their lives. This information was presented to decision makers and informed Vietnam's 2014 Law on Environmental Protection, with the inclusion of a child rights approach (UNICEF, 2015, p. 30). Similarly, in Zambia the 'Unite4Climate' strategy encouraged CaYP to become climate ambassadors which has resulted in the development of a range of initiatives to ameliorate the effects of the changing climate. Low-cost community projects such as a floating school in the Mongu area, is addressing a part of the country that is susceptible to flooding which is having a detrimental impact on school attendance (UNICEF, 2015, p. 34). Currently 1,000 child ambassadors have engaged at both government and local community levels, and through this have been able to work with more than one million community partners, offering a ringing endorsement of "the valuable role that CaYP can play in finding innovative solutions to issues that affect their daily lives" (UNICEF, 2015, p. 35).

Opportunities in the UK

At a national level there are a number of initiatives available to educational establishments to support sustainability in both knowledge and practice. These include: energy conservation; switching to more sustainable sources; developing wildlife areas for birds and insects within their local environments and installing solar panels. Some specific examples include:

- The 'Green Schools Project' (2024) which is working with schools to help them respond to current climate challenges, through teacher training sessions and support in reducing carbon emissions. Schools are encouraged to sign up and will then have access to a coordinator guide and information on activities to engage with pupils, including youth activism. This includes the Zero Carbon Schools Programme, which is part of the Green Schools Project. Feedback has demonstrated the impact this has had on school communities with one headteacher identifying that "Our pupils didn't fully understand what was happening with climate change. Now that they understand and they know what the issues are they've been much more conscientious about the environment" (Green Schools Project, 2024).

Other initiatives include

- Let's go zero (https://letsgozero.org/). A national campaign that encourages teachers, pupils and parents to work together to achieve zero carbon by 2030
- 'A Good Thing' (https://www.agoodthing.org.uk/). An App that matches businesses who have surplus resources to donate to charities who could make use of these.
- 'Solar for Schools' (https://www.solarforschools.co.uk/). This aims to educate the next generation on energy and decarbonisation.
- 'Learning through landscapes' (https://ltl.org.uk/). Support for different projects that recognise the benefits of outdoor learning and time spent

outside. Recent projects include helping schools to make their environment climate ready.

Community engagement

Case study of how one East Devon village went about making a difference

In 2021, in the middle of Covid, Chardstock EcoGroup in East Devon was formed by four residents who were concerned about their local environment and some of the challenges faced by the village and its environs. Lockdown provided the luxury of time and the opportunity to explore this further. Initially, a notice in the monthly newsletter brought concerned villagers together via emails and Zoom meetings. Later, as lockdown restrictions relaxed, face-to-face meetings took place in a greenhouse where hay bales ensured social distancing, while discussions explored what could and could not be achieved during Covid restrictions. Word continued to spread via the village newsletter and, as membership of the group increased, more projects were explored. Membership reflected all ages within the village, including parents, grandparents and CaYP.

Three initiatives were initially chosen to focus on including litter-picking which now covers allocated routes within the parish on a monthly basis. This was the first project to be instigated as it could be completed during Covid restrictions, either working solitarily or in pairs maintaining social distancing. For young people taking part, additional risk assessments were completed with the proviso that they needed to be accompanied by a parent. Rubbish collected was dropped off at one member's house, where it was sorted into different categories for recycling and, following contact with the local council, a separate wheely bin was provided for the collection.

The second area of interest concerned the environment and was in response to localised flooding. As the community experienced the impact of climate change close to home, it resulted in the local River Kit being identified as a specific project focus. Several challenges were facing the river as a result of climate change but, following research and liaison/communication with local authority groups, the EcoGroup decided to focus on river quality, and participated in a river monitoring system supported by West Country Rivers Trust (2024a). As part of the Westcountry Citizen Science Investigations (CSI) project, they became involved in monthly monitoring of the river water, measuring and recording water samples. CaYP were encouraged to explore the animals, amphibians and insects that inhabit the river and local area and recognise the importance of the water health on this. This resulted in joining the East Devon Catchment Partnership as the only community group, leading onto further opportunities.

This pioneering group became involved in identifying local issues and how these might be tackled, with the intention of reconnecting people with land and nature. The initial plan considered the Kit Brook catchment area field-by-field, river-reach by river-reach, recognising this was ambitious but self-determined by the people who live there. There was some financial and technical support, and they received funding to 'look, listen and learn' which resulted in a £10,000 grant (from a mixed blend of central and local authorities, private grant-making bodies and environmental charities), to engage in physical interventions to address one of the key issues affecting the area. As a result of climate change increased rainfall has seen the river water levels rising swiftly and, because of farming practices, the speed of this results in flooding as rushing water flows over compacted fields, taking soil with it. 'Leaky Dams' (West Country Rivers Trust, 2022) is one initiative that was, and still is, the focus of this intervention to address river flooding. Made of natural materials, they are a form of natural flood management (NFM) and are constructed at 'crunch points' of the water course, not the main river. In addition to slowing down the water flow, leaky dams also trap sediment, improving water quality further downstream. At the same time the dam materials are key to creating new habitats for river inhabitants and increasing biodiversity. Key to this was the involvement of local landowners and farmers, working in partnership to identify the most advantageous places/areas for the leaky dams to be constructed. This was not without its challenges as ditches need to be dug to provide an alternative flow for the excess water. The funding received supported employment of an expert team including a farm advisor and river technical expert and ecologist. To date the group have created 18 dams which take between five and ten years of continuous monitoring before they are fully established and self-maintaining. Additionally, this initiative has opened up dialogue between the group and local farmers, providing an opportunity to discuss the challenges faced by both sides. Many farmers are facing extremely challenging times and are keen to explore new methods of sustainable farming, but at times this is not helped when they receive more subsidy for ploughing up wildflower meadows than retaining them (Plantlife, 2024). Landscape recovery is the only way that river quality will get positive results, and evidence of landscape recovery projects funded by DEFRA (2023) demonstrate how successful these can be. Recent local flooding means some residents, both young and old, have experienced the impact of flooding on their local environment. It is hoped that CaYP will become involved in this ongoing project moving forward, as an extension of caring for their community.

The third project has been to work with a local primary school. The school was already committed to supporting children in developing a sustainable ethos including the development of a wildflower area. The school's Eco Club, supported by the village, have now planted an area of yellow rattle seed in preparation for a wildflower area. Yellow rattle is recognised as being particularly effective in reducing ground grasses and providing the most favourable conditions for the introduction of wildflowers (Save Our Magnificent Meadows, n.d.). In addition, the Eco Group have helped in the planning and building of a hibernaculum and bug hotel which has encouraged the CaYP to be more caring, and to further understand, about the food chain and the importance of protecting all wildlife in support of this. The hibernaculum was a particularly ambitious project and involved digging a deep hole which was then filled with both natural and manmade materials in order to create a place of safety below the frost line for hibernating wildlife. The original soil was then used to make a bump and overplanted with wildflower seeds. Wiltshire Wildlife Trust (2024) has provided a useful guide if you would like to consider building your own hibernaculum.

The children have also been following the life cycle of sea trout and over the past two years have looked after and monitored eggs in the classroom, before overseeing their release into local rivers. Fifty fertilised eggs were taken to the school and over two months the EcoGroup went in daily to check on progress and remove any casualties, culminating in a trip to the river where the children released them back into their natural habitat. This helped the children to connect with their local environment and understand more about the food chain and the importance of protecting all wildlife in support of this.

The group are also committed to reducing the use of plastic and as part of this commitment ran a session on Plastic Pollution "What's the problem with plastic?" with Key Stage 1 pupils. Children were clearly engaged in the topic, knowledgeable of many of the issues, and enthusiastic when contributing to discussions around the impact of plastic pollution and what they could do to help lessen its use. The session ended with children making a commitment to one activity that they could change, that would make a difference. Feedback from parents highlighted that their children had come home from school eager to share what they had learned. Engagement with the school has continued with regular weekly visits supporting its application for the Wildlife Garden Award, making bird feeders, creating bug hotels, creating a wild meadow area and making seed bombs, culminating in a week of activities devoted to engaging with the Kit Brook River.

Most recently, the EcoGroup organised a community event, 'Nature's Table,' to share the wider work of the EcoGroup and saw 300 visitors visit nearly 20 stalls and displays all focused on celebrating nature and how to protect it. This was an opportunity to share the wider work of the EcoGroup, while supporting Devon Wildlife Trust in distributing 300 trees. In addition to the EcoGroup's individual projects, other environmental groups were invited to participate such as the local plastic free group, insect and bee conservation groups and forest school representatives. Local farmers provided hay bales and children had the opportunity to make their own seed bombs, build their own bug hotels and take part in a 'wildlife' search. Two of the group's younger members had a stand where they shared the work they had been engaged in as part of their Duke of Edinburgh award, including the building and distribution of bird feeders and bird houses among the community. A number of local environmental groups attended and the success of the range of initiatives, and engagement with the community, was encapsulated in one visitor's remark stating that they were not surprised by the 'buzz' of the session and ending with "Well, you are known as the EcoHub!" On reflection, it was affirming to see that our work in the local area was important and that together, old and young, we can make a difference. This culminated in the group reaching the finals for BBC Radio Devon's Make a Difference Awards (2024) in the Green category, which recognises those who have had a positive impact on their local environment.

Other projects developed/promoted by the EcoGroup

- Becoming the first Devon community to be awarded a Wilder Communities Award in 2023 (Devon Wildlife Trust, 2023).
- Becoming a 'Plastic Free' community ally of Plastic Free Axminster, under the umbrella of Surfers against Sewage. The 'Talking Rubbish' event in 2022 organised by Chardstock Eco Group provided an opportunity for residents, including CaYP, to engage in a range of educational and fun activities while learning about waste, plastic and rubbish.
- Providing recycling opportunities for tablet blister packs and plastic plant pots.
- Supporting two students with their Duke of Edinburgh award which resulted in the making of bird feeders situated at different points around the village and fundraising to keep these filled; and the making of bird houses for distributing locally. A successful example of CAYP making a difference to their local community whist supporting the natural environment.

Personal commitments

As we near the end of this book it is a good time for us all to (re)consider our own contribution to some of the issues that have been raised and what we can do to lessen our own personal impact. CaYP can also be encouraged to take part, while adults can provide strong role models for our future leaders. A good starting position is to understand our own carbon footprint which can be checked using one of the many on-line calculators such as the WWF Carbon Footprint (https://footprint.wwf.org.uk/) and Carbon Independent (https://www.carbonindependent.org/). The questions can help us to understand how we ourselves are contributing to climate change, while identifying some ways in which we can make our own contribution to reducing our carbon footprints.

A few simple steps that we can all make include:

- Not buying bottled water. Bottled water produces 202.74g of CO2 emissions per litre, similar to driving your car for one kilometre! Fifty-two percent is as a result of the production and recycling of the bottles, with distribution, transport, and refrigeration making up the remaining emissions. In comparison tap water accounts for just 0.35g (Tappwater, 2024).
- Reducing the environmental impact of your diet. The United Nations (nd) Climate Action site listed here includes useful information and suggestions for exploring changes that you might want to consider: https://www.un.org/en/climatechange/science/climate-issues/food.
- Signing up to The Big Plastic Count (https://thebigplasticcount.com/) that takes place in March each year. By counting your plastic waste for one week, you can better understand the amount of incidental plastic that your household generates in a single week. Once you have uploaded your results you will receive your household's plastic footprint, together with advice on how to reduce your plastic. Your results will then be used to calculate the national results which will be used to lobby the government for decisive action particularly discussions around the Global Plastics Treaty referred to earlier.
- Finding local groups that you can join. Examples of groups with branches around the UK include: Friends of the Earth (https://friendsoftheearth.uk/), Surfers Against Sewage (https://www.sas.org.uk/) and Plastic Free Communities (https://plasticfree.org.uk/). Or start your own group as Chardstock EcoGroup did.
- Sourcing sustainable fashion: clothes have a high carbon footprint. Follow the mantra 'Reduce – ReUse – Recycle.' Buy fewer clothes, use many of the second-hand shopping apps that are easy to use to buy and sell your own clothes. Hold clothes swaps and check whether your favourite retailer has a swap option. Many now provide voucher incentives in exchange for your old clothes.
- Creating handprints. Biemer (2021) encourages us to create a past and a future handprint by drawing an outline of your hands and on one writing what you have already done for the environment, and on the other what

you are going to do. Use each finger to identify a specific thing you have done or intend to do. Treat it as your manifesto, that you can add to, revise and revisit over the course of the years.

Conclusion

This chapter began by reflecting on the world that our youth and future generations will inherit, and the impact this is having on their well-being. It then celebrated the activism of young people around the world who have become climate change activists, putting increased pressure on the world's governments to make changes. This activism has provided an outlet for young people to manage their anxiety about the future and through this provide a more optimistic view. An important role for educators is to build this capacity to cope with a world that is altering as a result of climate change and see this as an opportunity as much as a threat. There are a range of national initiatives and local groups that are eager to work with schools and support this process, giving our children and young people agency in the shaping of their world. We have considered initiatives which inspire hope at global, national and community scales. The Chardstock EcoGroup case study was an uplifting example of how a small group of people in a village got together during Covid to form a community group which started tackling small eco issues, before expanding to become a thriving group. Finally, there was an opportunity to encourage each of us to reflect on what changes we could put into place, to make a difference ... nothing is too small.

To conclude, as we come to the end of this chapter and book you are encouraged to complete your own 'handprint' and consider what changes you could make, or support others to make, when working towards a more sustainable world. The sustainability mantra of 'Reduce-recycle-reuse' is a good place to start. Write a letter to yourself, or your children, or future generations and tell them what you are going to do to protect their world. As the renowned zoologist and primatologist Jane Goodall said, "You cannot get through a single day without having an impact on the world around you. What you do makes a difference, and you have to decide what kind of difference you want to make" (Jane Goodall Institute, n.d.). Together, we can all make a difference.

References

Biemer, J.R. (2021) *Our Environmental Handprints. Recover the Land, Reverse Global Warming, Reclaim the Future*. London: Rowman and Littlefield.

Big Plastic Count (2024) The big plastic count. Available at: https://thebigplasticcount.com/ (Accessed 5 March 2024).

Chardstock EcoGroup (2024) Facebook. Available at: https://www.facebook.com/groups/chardstockecogroup/ (Accessed 24 February 2024).

DEFRA (2023) Landscape recovery: Sharing the successful second round projects. Available at: https://defrafarming.blog.gov.uk/2023/11/29/round-two-projects/ (Accessed 24 February 2024).

Devon Wildlife Trust (2023) Wilder communities award. Available at: https://www.devonwildlifetrust.org/wilder-communities-award#:~:text=What%20is%20the%20Wilder%20Communities,award%20to%20celebrate%20your%20efforts (Accessed 24 February 2024).

Gilbert, P. (2015) Inherit the earth, or borrow it from our children. Available at: https://www.nrpa.org/blog/inherit-the-earth-or-borrow-it-from-our-children/#:~:text=The%20report%20starts%20with%20the,generations%20runs%20throughout%20the%20document (Accessed 20 October 2024).

Green Schools Project (2024) Zero carbon schools. Available at: https://www.greenschoolsproject.org.uk/

Hickman, C., Marks, E., Pihkala, P., Clayton, S., Lewandowski, R.E., Mayall, E.E., Wray, B., Mellor, C. and Van Susteren, L. (2021) Climate anxiety in children and young people and their beliefs about government responses to climate change: A global survey. *Lancet Planet Health* 2021(5) 863–873. Available at: https://www.sciencedirect.com/science/article/pii/S2542519621002783?ssrnid=3918955&dgcid=SSRN_redirect_SD (Accessed 29 July 2024).

Jane Goodall Institute (n.d.) What we do makes a difference. Available at: https://janegoodall.ca/what-we-do/ (Accessed 20 October 2024). What We Do | The Jane Goodall Institute of Canada.

Plantlife (2024) A big win for grassland, but farmers need more. Available at: https://www.plantlife.org.uk/our-work/a-big-win-for-grassland-but-farmers-need-more/ (Accessed 20 October 2024).

Sanson, A.V., Van Hoorn, J. and Burke, S.E.L. (2019) Responding the Impacts of the Climate Crisit on Children and Youth. *Child Development Perspectives* 13(4):201–207.

Save Our Magnificent Meadows (n.d.). Managing Meadows - Plantlife. Available at: https://www.magnificentmeadows.org.uk/assets/pdfs/Using_yellow_rattle_in_restoration.pdf (Accessed 4 August 2024).

Tappwater (2024) What is the carbon footprint of bottled water? Available at: https://tappwater.co/blogs/blog/carbon-footprint-bottled-water#:~:text=Based%20on%20the%20most%20cited,592%20km)%20with%20a%20car (Accessed 30 July 2024).

Thunberg, G (2021). Keynote at Youth 4 Climate Pre-COP26 (speech). Youth4Climate conference, Milan, 29th September. Available at https://www.youtube.com/watch?time_continue=228&v=n2TJMpiG5XQ&embeds_referring_euri=https%3A%2F%2Fwww.bing.com%2F&embeds_referring_origin=https%3A%2F%2Fwww.bing.com&source_ve_path=Mjg2NjIsMTM5MTE3LDI4NjY2 (Accessed 25th March 2025).

Thunberg, G. (2022) Available at: https://earth.org/greta-thunberg-quotes-speeches-to-inspire-climate-action/ (Accessed 29 July 2024).

UN (n.d.) Climate action. Food and climate change: Healthy foods for a healthier planet. Available at: https://www.un.org/en/climatechange/science/climate-issues/food (Accessed 4 August 2024).

UN (2023) The 17 goals. Available at: https://sdgs.un.org/goals (Accessed 6 March 2024).

UN (2024) UN News. Global perspectives Human stories. Available at: https://news.un.org/en/story/2024/02/1146737 (Accessed 1 August 2024).

UNDP (2024) Loss and damage fund for developing countries. Available at: https://www.undp.org/belarus/stories/loss-and-damage-fund-developing-countries#:~:text=Named%20the%20%22Loss%20and%20Damage,brunt%20of%20climate%2Drelated%20challenges (Accessed 5 March 2024).

UNICEF UK (2013) Climate change: Children's challenge. Available at: https://downloads.unicef.org.uk/wp-content/uploads/2013/09/unicef-climate-change-report-2013.pdf (accessed 31 July 2024).

UNICEF (2015) Children and the changing climate. Available at: https://www.unicef.org.uk/wp-content/uploads/2015/11/Unicef_2015childrenandclimatechange.pdf (Accessed 29 July 2024).

Watts, J. (2019) Greta Thunberg, schoolgirl climate change warrior: 'Some people can let things go. I can't' Available at: https://www.theguardian.com/world/2019/mar/11/greta-thunberg-schoolgirl-climate-change-warrior-some-people-can-let-things-go-i-cant (Accessed 20 October 2024).

West Country Rivers Trust (2022) Leaky dams help slow 'leaks' in Plymouth. Available at: https://wrt.org.uk/leaky-dams-help-slow-leaks-in-plymouth/ (Accessed 24 February 2023).

West Country Rivers Trust (2024a) Bringing rivers to life for 30 years. Available at: https://wrt.org.uk/ (Accessed 24 February 2024).

West Country Rivers Trust (2024b) Be a citizen scientist with us. Available at: https://wrt.org.uk/westcountry-csi/#:~:text=Our%20Westcountry%20CSI%20aims%20are,with%20as%20quickly%20as%20possible (Accessed 24 February 2024).

Wiltshire Wildlife Trust (2024) How to build a hibernaculum for amphibians and reptiles. Available at: https://www.wiltshirewildlife.org/hibernaculum (Accessed 5 August 2024).

CONCLUSION

Nicola Watson and Carla Solvason

The introduction to this book was written on a misty autumn day in the UK. As we conclude, we are surrounded by a beautiful layer of crisp winter snow, but we know that our seasons are changing and becoming increasingly unpredictable; in some areas of the world, with devastating effect. This book is not intended to incite fear, as fear debilitates, but it is intended to invite reflection on the role of the educator in the 21st century. This requires us to acknowledge the uncomfortable truth that we live on a fragile planet with finite resources and that our rate of consumption cannot be sustained. This acknowledgement can result in eco-anxiety, something which can affect any one of us. Educators play a key role in shaping the future, the recognition of which is potentially daunting. Succumbing to the temptation to carry on 'business as usual' is we suggest, a form of denial and as such, a symptom of fear and debilitation. Reframing educators' roles as pedagogues, using their talents and resources to create opportunities to improve the lives of others, enables intrinsic reward through engagement with purposeful and meaningful endeavour. This contributes to good mental health for educators as well as those with whom they work.

By reconsidering our mindsets and our values, acting as role models for the CaYP who take their lead from us and advocating for the most vulnerable, we can better align our actions with our values. What should we be demonstrating in our conversations and our day-to-day actions? Reading through the chapters of the book you will recognise some key concepts that emerge again and again:

- Our planet is fragile and is buckling under the thoughtlessness and heavy-handedness of the humans inhabiting it. We need to re-establish a caring relationship with our planet.

DOI: 10.4324/9781003506850-17

- We are all in this together. Despite the numerous traits which make us different, we all share stewardship of the Earth. Every one of us needs to acknowledge this responsibility, and work with one another to enable the planet to heal.
- We need to reject the neoliberal mantra that 'more is better' and nurture our appreciation of 'enough.'
- By sharing our concerns, working with others and advocating for CaYP, together we can help hold our policy-makers to account. Through social and political engagement, we help shape policy which focuses on the long-term health and wellbeing of our world rather than acquiescing to policies focusing on s short-term electoral goals.

  One statement stood out in this book for us. It came from Sam Jackson in Chapter 11 who said:

  implementing a global change on this kind of scale will only come about through a truly collective action, a sort of coming together to achieve communal ends – which I believe, when we zoom in a bit, is weaved out of the fabric of interpersonal and inter-community relationships.

If ever there were a statement to sum up 'we are all in this together' then it is this one. We hope that this book has inspired you in some way to reflect upon your own context. Each of us has our own unique inner-resources, such as creativity, imagination, intellectual, spiritual and practical skills which can be garnered and shared so that together we can help to shape a sustainable future for ourselves, our children and the generations to come.

INDEX

Note: **Bold** page numbers refer to tables.

Kolibri 360 GmbH
Ferdinand-Jühlke-Straße 7
99095 Erfurt, Deutschland
produktsicherheit@kolibri360.de